P9-CSH-273

"The Warrior Writers workshop cultivates a needful dialogue among these veterans amidst the process of confronting themselves as a means of returning from the war."

–Yusef Komunyakaa
Vietnam Veteran and Pulitzer Prize winning poet

"Gruelingly honest, hones like a flute atop a mountain at the fatty tissues around the heart and expresses the singular and most chilling music of truth about what a human being must endure—and yet these poems are beautiful, harken back to the time of sincerity and compassion and understanding. This anthology should be a must read for high school kids and college students and should be read by the family, every business person, should be the clarion at dawn in our lives that raises our awareness to what we are allowing to happen to our brothers and sisters."

–Jimmy Santiago Baca
American Book Award and Pushcart Prize winner

"The poems (and images) in this anthology are to begin with, the most important poems of witness you may ever read. They offer a counter-narrative seldom allowed us in the mainstream conversation about war—the hearts and minds of men and women we send off to unspeakable horrors in the supposed name of freedom. And because the lines are so clear, the works so pregnant with the complicated knowing of both what is said and unsaid, we can't look away. Support truths that others would have hid from you. You'll be a better person for the effort."

–Roger Bonair-Agard
National Poetry Slam winner

"Combat Paper and Warrior Writers have been lifelines that stand in the way of avoidance, isolation, depression, and suicidal ideation."

–Sarah N. Mess
Somalia Veteran and Warrior Writer

"Participating in Warrior Writers has been invaluable to me. I have found a fellowship of understanding and a sense of community. It has been a supportive space for me to express my creativity and find my voice."

–Patrick Majid Doherty
Iraq Veteran and Warrior Writer

Copyright © Warrior Writers, 2014. All rights reserved.

All poems, prose and artwork appear by permission of the respective artists. All rights revert to the respective artists. No part of this book may be reproduced, in any form, without written permission from the publisher. Email requests to: info@warriorwriters.org.

Warrior Writers is not responsible for content of any individual piece.

Justin Cliburn, "The Trauma of Lost Love" originally appeared on The Good Men Project blog.

Preston Hood, "Crossing a Minefield" is a different version of "Against Fear," originally published in *A Chill I Understand* (Summer Home Review Press, 2006).

Nathan Lewis, "How to Make a Combat Paper Book" and "Diesel Truck Time Machine" from *Colors of Trees We Couldn't Name* (Combat Paper Press, 2013).

Hugh Martin, "Memorial Day" was originally published in *The Iowa Review*, Spring 2013 (43.1); "First Class Guilt" originally appeared on therumpus.net (2013); "Home From Iraq, Barking Spiders Tavern" and "The Stick Soldiers" from *The Stick Soldiers* (BOA editions, 2013).

Peter Sullivan, "The Weapon" and "I Decline" from *Carry This For Awhile* (self-published, 2013).

Brian Turner, "The Baghdad Zoo" from *Here, Bullet*. Copyright © Brian Turner, 2005. Reprinted with the permission of The Permissions Company, Inc. on behalf of Alice James Books, alicejamesbooks.org.

Bruce Weigl, "Song of Napalm" from *Archaeology of the Circle: New and Selected Poems* (Grove Press, 1999).

Printed by L. Brown and Sons Printing, Inc.
14-20 Jefferson Street
Barre, VT 05641

ISBN 978-0-692-22210-2

Edited by Lovella Calica & Kevin Basl
Layout and Type Design by Rachel McNeill
Cover Art by Aaron Hughes

WARRIOR WRITERS

A COLLECTION OF WRITING & ARTWORK BY VETERANS

Edited by
Lovella Calica & Kevin Basl

This book is dedicated to Joshua Casteel.
Your passion, commitment to justice, creativity and faith
are a torch of inspiration to all of us.
You were taken too soon, but your spirit will shine on.

December 27, 1979 - August 25, 2012

WARRIOR WRITERS

Our mission is to create a culture in which veterans and service members speak openly and honestly about their military experiences. In the form of veteran-focused writing and art-making workshops, we provide a creative community for self-expression and reflection, while also fostering mutual understanding and peer-to-peer support. Participants are encouraged to share their art in the form of books, performances, and exhibitions, thus empowering veterans to tell their stories and voice their opinions in front of a wider audience. This process also provides opportunities for the public to gain a deeper sense of the lives of veterans and service members—something sorely needed in a society where so few serve in the military.

Although Warrior Writers is a program of Culture Trust Greater Philadelphia, members and workshops can be found across the United States. Warrior Writers is not part of or funded by the Department of Veterans Affairs or any therapy programs. It is important that we maintain our autonomy because some veterans are resistant to institutional settings; many have said they feel comfortable with Warrior Writers because we are informal and provide a real sense of community. The organization started organically (in 2007, Lovella Calica was inspired to lead a writing workshop for her veteran friends after several of them shared their poetry with her) and that same grassroots energy continues to drive us.

From the beginning, we've partnered with a diversity of people and institutions. We've worked with libraries, hospitals, homeless shelters, museums, coffeehouses, and theaters. From Walter Reed patients to student-veterans, from active duty soldiers to members of Iraq Veterans Against the War, we've worked with many veterans of varied political affiliations. We've read poems in front of hundreds of opera-goers at Opera Philadelphia, and workshopped with Pulitzer Prize-winning authors. We've consulted for university writing programs; we've helped veterans who could barely hold a pen. Warrior Writers performances have been hosted by the Nuyorican Poets Café, Live at Kelly Writers House, USOs, bookstores, galleries, and more.

We're also honored to continue working with our friends at the Combat Paper Project (some of the visual artwork contained in this anthology is printed on Combat Paper). Founded shortly after Warrior Writers by artist Drew Matott and soldier-turned-artist Drew Cameron, the Combat Paper Project conducts workshops around the world teaching veterans and their communities how to turn military uniforms into handmade paper—an excellent medium for sharing their stories. Warrior Writers looks forward to working with the Combat Paper Project and its affiliate paper mills for years to come.

the first time I wore a uniform it
had a white paper tag attached with the
military service number, size and description

[Medium Regular]

I never really felt much of either
but the barcode explained otherwise

[Thin Casual]

[Young Uncertain]

DO NOT REMOVE THIS LABEL

Photo by Lovella Calica

CONTENTS

EDITOR'S NOTES

The work that fills this book took months of encouragement (both from me and others in our community) and rewriting on the part of the authors. To say that the material did not come easy is an understatement. Consider the trauma, grief, loss— the sources of much of this writing. I ask veterans to confront their fears, the haunting faces, the weight of their military experience. I ask them to sit with their pain, to turn it on its side, to remember, and to re-create themselves on paper.

Of course, it's not as simple as writing it once then walking away. Chantelle Bateman's poem about the frustration of her VA compensation hearing started out as just a few lines scribbled in her journal. Often, we discussed her desire to articulate the absurdity and invasiveness of the claims process. I gave her feedback and made suggestions on several versions of the poem. "I don't want to go back to that claustrophobic room," she would say when I asked how revisions were going. I reminded her that she wanted to tell her story, that she would be proud when it was finished. A month later, she emailed me a draft of the poem. I was ecstatic—she nailed it! Revisiting her difficult experience, she used creativity to turn it into a piece of artwork that (at the risk of sounding bold) has the potential to reshape both herself and the world around her.

Another piece included in this anthology came about last year during a weeklong workshop in Washington DC, working in conjunction with Combat Paper NJ. Alex Fenno, a Marine, wrote a letter entitled "Sorry for not being Sorry" to some street-children he had encountered in Afghanistan. When it came time for us to compile this book, I asked him to submit the letter, requesting that he work with us on revisions. He was eager to dig in, excited to continue writing and engaging with us. In the new version of his letter, Alex writes that he has spent a lot of time reflecting. Through the revision process, he was able to explore new perspectives, coming to deeper realizations about himself that may not have happened had he never returned to the work.

Unfortunately, there is a lot of material that never made it into this book. Maybe a journal was lost, or the writer didn't have the emotional energy to return to his or her work. I recall a poem, "Progress," that Warrior Writers artist Leonard Shelton wrote last summer in a workshop at Antioch College. In the poem, he shares that he's finally learning how to enjoy the 4th of July fireworks again, with the companionship and love of certain people (and animals) in his life. To some this may seem insignificant—watching fireworks—but this is indeed progress, both as a veteran overcoming post-traumatic stress symptoms and those civilians intentionally engaging in the re-integration process, supporting our veterans in genuine and tangible ways. Though the poem was lost, the pride that Leonard expressed during that workshop remains with me today.

After 10 years of working with veterans, I am still learning new things, witnessing growth and overcoming the challenges of working with veterans. I have learned a lot about how post-traumatic stress manifests, how it affects relationships, and how it ripples through our society. Spending much of my life working through my own childhood trauma has helped inform my understanding. Writing is one tool that has given me a space to process the past. It has been an honor to share this tool with my peers, to witness its positive effects again and again. Though this road of confronting ghosts and sharing stories is exhausting, I remain on it, knowing our lives are bound together, that we share these struggles as a society, and that art can provide a path forward. We share songs and laughter, watch our children grow, and together hope for a more loving, creative world. I continue to love the men and women who make up this community, and remain committed to finding more people willing to take this walk with us.

* * *

Writing a Decade

15 years ago, I would never have believed it
10 years ago, 5 years ago, I just met you

you had been angry
drinking at bars, alone
fighting, crying, calling at 2 am

slept with a gun at night
were spilling pills, powders
couldn't leave your house, wouldn't answer your phone

you were pacing, shouting, distrustful, resentful

we sat together
talked, wrote, cried, yelled, laughed
fought, stood up, found we had voices, used them

learned to look up, look around
learned to lean on each other
set our feet more firmly on the ground

your body is no longer something to punish
shedding habits you learned long ago
you take it outside, go fishing
stretch, do yoga, dance, run, strum
fill it with nutrients instead of chemicals
you have a vision now, focus, purpose, confidence

we've been walking together for years
still digesting the mess
molding it into more

I see you loving
I see you struggling
I see you smiling
still crying and trying
see you putting yourself together

I see you speaking up, I see you growing
I see you getting stronger, smarter
readjusting, recalibrating
I see we all have more work left to do

we'll keep walking together
learning, trying, helping, growing
cheering each other on along the way

– Lovella Calica, Founder

FOREWORD

On the Temple University train platform, I listened to a fellow classmate tell me about a writing organization for veterans based out of West Philadelphia. Warrior Writers? She knew I was a vet from the stories I had been submitting to our creative writing workshop and suggested I check it out. I appreciated the gesture, but wasn't interested. One year into my fiction MFA, I was certain I was getting all the tools needed to process the struggle, rejection and (occasionally) the small successes of the writer's life. To me, writing was a solitary activity, best done at a secondhand desk in one's mice-infested studio apartment. What this arts organization had to offer surely didn't apply to me. Furthermore, I didn't want anything to do with the military or veterans' organizations in general. My Army enlistment had ended in 2008, following a year-long stoploss, for which I deployed to Iraq for a second tour. The Army then served me muster orders in 2009, saying I would possibly be recalled to active duty, to go back for a third tour, if I didn't join the Reserves. After the Army finally did let me go, I was so soured on all things DOD that I wanted to get as far away from military culture as possible. That also meant veterans.

For two years, I attempted to turn off my former identity, the one dictated by rank and marching step. I told a select few I was a veteran, and to those whom I did tell, I gave no specifics. But as I've come to learn from more than a couple Vietnam vets, you can't just turn off war. Throughout my silence, I could sense something was missing, something I couldn't articulate. Classmates surrounding me felt foreign, as if I were an exchange student experiencing a language barrier. I thought I couldn't share my military experiences with my peers because, aside from the age gap, we had no common ground on which to discuss war and its consequences. My favorite activity became spending long hours in my apartment, alone (it was the writer's life, remember). What I would come to discover is my self-identity had changed dramatically through five years of service, and I was still trying to live as my old self, the me before the Army. "Writing around" things I had experienced in Iraq through fiction—attempting to write about the "new me" in the third person—meant my most troubling stories stayed locked away in memory, that I wasn't claiming ownership of them.

When I finally did come to Warrior Writers (a mutual friend introduced me to Lovella) I discovered what I had been lacking: a genuine, face-to-face community where I could share my stories honestly, a place where I could understand my new identity. I found myself surrounded by friends to encourage me both in my art and in processing my war experience, to help rekindle a sense of purpose. This was something essential that no MFA program could provide.

Warrior Writers is not a panacea, of course. Our workshops are not "art therapy." Rather, we are a community of artists (who happen to be veterans) sharing our stories, reevaluating the past, challenging popular misconceptions of war, and--with more than a little hard work--aspiring towards great art. Writing is inherently therapeutic; that's a given.

Although ours is a tightly-knit community, this book, our fourth anthology, was not easy to publish. Collecting the material, revising one-on-one with the authors, and, on a more personal level, encouraging the emotional and psychological digging necessary to produce such private and (often) painful work was no overnight task. In my own experience, I doubt I would have ever talked publically about the sources of two of my poems included herein: "Wake Up Call" stemmed from a violent suicide that happened in my barracks just prior to deployment in 2005; "The Noise Remains" is my attempt to communicate, in the confessional tradition, what might have gone on behind one veteran's wall of silence. The community encouraged me to share, and I did.

Some of the material in this anthology is the spontaneous and raw outpourings of first-time creative writers; some is the result of years of polishing and craft development. Regardless of writing ability (we'll work with you), education, service era, or circumstances of discharge, Warrior Writers is open to all veterans. The work herein has been interspersed in the same way participants might be seated around a workshop table, in dialogue. To place Vietnam veterans alongside those of the Gulf War, Somalia, Iraq, Afghanistan and other recent conflicts is a truly unique experience, and the knowledge that gets generated and shared, talking and writing as equals around the table, has proven invaluable to nearly all who have participated. Lieutenant colonels sit alongside privates; Marines and airmen discuss what it felt like to put on a military uniform for the first time. Beyond race, gender, sexual orientation—at our table, veterans and service members trade stories in confidence. This anthology provides a glimpse into that very distinct and vital place.

– Kevin Basl, Co-Editor

Photo by Willie Young

THE WEAPON

I know about the battle going on inside your head
It's dark. It's late. The memories that follow you to bed
The time has come for you to choose how this war will end
Continue taking casualties or try brandishing a pen

Your weapon is the ink fired through your fingers via tears
Fragmentation sentences to neutralize your fears
High explosive tracer words dispatched a thousand at a time
Daisy-chained together setting paragraphic landmines

The only way you'll ever win this war or have a chance
Aim your pen. I've got you covered. Attack! You must advance
Fight this battle hard enough, and you'll be tired but you'll see
The ink you're using as a weapon becomes an instrument of
  peace

Photo by Lovella Calica

BEFORE MORNING

There is something compelling about the dark
before morning. The still silence, the stark

contrast to the violence of day masked
in the predawn. Eight years ago, when tasked

to assault the city and liberate
those people, a hesitation to break

that quiet lingered until night was breached
by the engineers, and the assault reached

the morning, while that morning lingers on
eight years later in November, before dawn.

STANDARD OPERATING PROCEDURE

Embark:
You will be told to forget your previous life–
       Your mother
       Your father
       Your siblings
       and home

You are in God's hands now

Your focus now is to kill
       To learn how to kill
       To teach others how to kill
You will become a professional
       trained in killing

Training your muscles to remember the
       Slow
          Steady
             Squeeze
       Between breath

Learning to disassociate as you pray
       so that only your body is present
       in dying–
             No one in their right mind
             would kill a man

Remember:
God matters now
He will forgive you for killing
He will keep you from being killed
He will take pity as you weep and watch
         others kill

God will let you remember what it was like to love
         But only for a moment
You are not home anymore
         You are in God's hands

GOD   COUNTRY   CORPS

GOD   COUNTRY   CORPS

GOD   COUNTRY   CORPS

You are in God's hands
         You are not home anymore
But only for a moment
         God will let you remember what it was like to love

He will take pity as you weep and watch
         others kill
He will keep you from being killed
He will forgive you for killing
God matters now

Remember:
No one in their right mind
                    would kill a man—

So that only your body is present
in dying
You will disassociate as you pray
            Between breath
Squeeze
Steady
Slow

Your muscles are trained to remember
That you became a professional
            trained in killing
                    Teaching others how to kill
                    To learn how to kill
            Your focus is to kill

In God's hands
Your previous life—
                    Your mother
                    Your father
                    Your siblings
                    and home

Embark

God   Country   Corps

GODS MAKE TERRIBLE GENERALS

Up there in front the commander
 gave his big game speech
Two days before the bombing started
"Going to war for God"

The Steel Warriors
Cold War weapon system boondoggle saints
A mean group that fires telephone poles
filled with cluster bombs
Atheists in the ranks squirmed like burning worms

We punched our biblical time cards
 amidst a missile attack in Kuwait
Through the sand storms, the terrifying house raids
All the lonely nights wondering the range of an old
 mortar tube
The greasy black stains on the road next to a flattened
 sedan
Through all that
We prayed

Some prayed to Gods
Others for blood and fire
I prayed to the angel of chance and circumstance
Kept my mind as low as possible
Sex, food, cars, football, sex
Some prayed to Gods

My dog tags read No Religious Preference
My Kevlar never felt like a Halo
My M-16 never smelled of blonde baby hair
 or Frankincense

Not Myrrh or ripe Cantaloupe
Not clouds or clean feet

It smelled of sausage and money
Spent old rolled up coke bills cut with cordite

In Iraq that same Commander told us
 to run over children for our own safety
"They'll use kids to stop the convoy"
God's war
Words like a 69' Camaro
 burnout brake stand of hypocrisy

Handing out food to the starving kids was forbidden
Who would Jesus throw a meal to?

Holy trinity in the burnt out tanks returning to sand
We passed remnants from Father's war
The son dispatched us
 without ever knowing the feeling
We met the ghost of Vietnam
It welcomed us with a grin and laughter
Back so soon?

Standing with an armload of tired old one liners
we swaggered drunk on power and war budgets
Armed, angry and motivated
Resplendent in schools, hospitals, roads
A salad of history's violent fools thrashing in the sand
following well worn and rutted roads
Imperialists, Crusaders, Idiots,
Conquistadors, Missionaries, Carpetbaggers
Shameless mercenaries at best

ARMY VALUES, TRANSLATED

Loyalty
Defer all decisions and moral obligations to an outside authority. Relinquish choice, reason, and responsibility for your actions.

Duty
Do what you are told when you are told without complaint or hesitation. Put your family, life, and health behind the needs of the Army.

Respect
Respect must be given to everyone that outranks you. Do not expect to be treated with respect. You have no power.

Selfless Service
Do not feel sorry for yourself, resist, or even think of taking care of your own needs. Your needs are irrelevant.

Honor
Honor the flag and the service. Do not honor yourself, your sisters in arms, or your fellow humans.

Integrity
Act as if you've been a part of a noble cause. Insist we've been trying to help. Do not express doubt in the mission.

Personal Courage
Endure more than any human can. Come back for more. Pretend it doesn't hurt.

YOUR GUILTY ENEMY

childhood prepared me for war,
when I arrived it became no surprise
what individuals could do, even to a child.
I belonged as the ten year-old-boy
cornered in his bedroom, terrorized
when the artillery shells began to fall.
I deserved to be the young man
ransacked and disappeared,
while his family condemned
to horror with death in darkness,
yelled at with unintelligible nonsense
not even meant to be understood.
I should have been the woman in the streets
with all the powers of the world
pointing down manly barrels,
unable to articulate why I deserve to live
why I deserve to be left alone.
I am the child throwing rocks
forced with punishment for the deeds of others,
or perhaps for what I have,
in failed hope, done in desperation for life.
I deserved to be the tortured soul
locked up in my own house
of forced degradation and ridicule
for daring to live, daring to defy
daring to believe the extent
of my freedom, my existence.
I am the skin colored with shame
forced to look at me from the outside,
and I can see how you sometimes
confuse me with innocent enemies.

WE'RE READY UNCLE SAM & CONTEMPLATION

The *We're Ready Uncle Sam* series, as well as *Contemplation*, represent the effects of violence that are portrayed and emboldened by our media, video games, and toys. In addition to numerous others, these outlets promote calloused views in our children. Whether it is in the form of simple war play or through our society's support of the country's military engagements, these images attempt to question how toys and the behavior associated with them promote violent behavior. Further investigation reveals a bewildering dichotomy: a distaste towards ferocious behavior, and a willingness to allow children to participate in such actions through simple play. Consequently, the nation's youth are encouraged to consider military service. My conscience is challenged by this idiosyncrasy, not only as a father, but as a humanist who inadvertently inspired his two boys to want to become soldiers.

We're Ready Uncle Sam I, II & III, pp. 35-37
Contemplation, pp. 38-39

NIKE BOY

You came to me with a teenager's grin
And sand-scratched eyes
And the Nike shoes flopping on your too-small feet
Gifted by the infantryman who captured you
And was moved by your bloodied, bare feet.

We huddled at the small table in the interrogation tent
As a sand storm raged outside
I touched your face as I washed your eyes
I wiped your tears as they ran down your dusty cheeks
You touched my heart as we talked
Not as interrogator to prisoner
But as two teenagers curious about each other.

The hell of war fell away
As you grinned down at your Nike shoes
And grinned up at me
A poor village boy, these were your first shoes
Like any teenage boy anywhere
You were delighted with your prize.

I asked you my interrogation questions
You answered, but then asked some of your own
About America, about music, about shoes
You, my enemy, were just a boy of 17
Just a boy.

Had we met in a café
We might have become friends
As it was, for that moment in time,
We let ourselves be
Just two teenagers, curious about each other.

IMPENETRABLE

They penetrated through a ballistic window,
Bored through two inches of armor,
Pierced the plate-carrier,
Plowed through a ceramic plate,
Tore a damp, dust covered ACU blouse,
Slit open a stink filled undershirt,
Sliced through salty, sticky skin,
Splintered a dense chest bone and embedded
 into my heart,
Sun-kissed, dirty cheeks brought gentleness
    to strikingly olive eyes,
Unblinking curiosity,
Void of any hint of sin,
Thank you little Iraqi girl,
You reminded me that beauty and innocence
 still live amongst the battered mess.

SEARCH & RECOVER

You want to reach out & touch them tell them
their mission will always fail because we
can see the heat of their bodies
radiating against cold night. You want
to reach out & turn them around, say go
home to your wife, go home to your children
but they come, the Euphrates their highway
in small boats lugging Kalashnikovs
to die on the river's now naked shore
line they die on the river's edge dovetail
rounds & shrapnel claymores & handgrenades
people say you never hear the bullet
that kills you but how would they know? These men
these Iraqi men, ghosts in thermal sights
knew something that I didn't that I'm just
understanding a decade after. Next
morning we searched for their bodies, maybe
in pieces maybe in the park where children
played soccer & we found nothing except burnt
ground.

MY RIFLE

Taking my rifle apart
Cleaning each piece
Watching them glisten in the sun
Like oil wrapped jewels
The fumes, sand and sun focus me
Fingers moving mechanically
My mind adrift yet knowing
That I can put it back together again
But that I can never put back together
What my rifle and I have already taken apart
Those pieces glistening in the sun
And in my mind
The fumes, sand and sun focus me
My innocence, my youth
Reflected in sightless eyes
My rifle and I move mechanically
As I clean and reassemble her
For the task of reassembling
Anything but my rifle
Is lost out there in the sand and sun glistening
Somewhere in my soul left for another day
Another world
Because today I am a soldier!
And tonight I sleep with my rifle!

ABANDONMENT OF AN M-16

I was supposed to be there
By her side no matter what
But she left me

The lectures, the speeches
She gave to soldiers about ensuring we were by their sides
Their trusty partners in this unpredictable war
But she forgot me, she left me

I lay there on cold concrete
Listening to the sounds of metal cutting through the air
She scrambled out there
Trying to save those who could not make it on their own

To the concrete sanitation ditch
That would have protected them
I was discarded as dead weight
As she crawled to grab them
Trying to pull them back into safety
I could have been helpful
But she left me

I could have shot back
As bullets flew by, grazing by their bodies
Cutting the air around them
Pinging off the metal and kicking up gravel rocks that
surrounded them

But she left me
I could have killed those bastards
So she didn't lie there
With her body hovered and draped over her soldier
Crying for his child
Wondering if they would be united again
She could have had me to protect them both
But she left me

She didn't have to be out there alone
She was trained to have me by her side
I was supposed to protect her
But she left me

She chose to go alone
And there I was
Just a hunk of metal, lying on concrete
With chaos all around
I was nothing without her
Maybe if she had taken me
But she left me

ANOTHER "BREAK 'EM DOWN, BUILD 'EM UP" FORM OF PUNISHMENT. WHAT IT WAS SUPPOSED TO ACCOMPLISH, I HAVE NO IDEA.

KILGORE©2009

BASIC TRAINING

I was primed to strut my stuff
In basic training–
I already knew the drill
From Boy Scout camp–
But I wasn't prepared for
The abrupt lesson
In abandoning fallen comrades–

"Listen up!" Sgt. Cutter,
Combat medic badge sparkling
On his chesty chest, shouted
One morning in formation–
"One of you dickheads slit his
Wrists last night–
Next lily-livered loser
Who wants ta slit your weak-ass wrists–
I'm handing out razor blades!"

A DAY LIKE ANY OTHER

I came to work, a day not unlike any other
Beautiful September morning, sky a crisp, crystal blue
The excitement of the Pentagon, soldiers doing what
  soldiers do
Actions and decisions being worked, meetings to attend
Another day of our Army at peace

A day like any other, but then it wasn't
A plane hits the Trade Center!
Once is an accident. Twice? I don't think so
Heightened awareness, but life goes on
Phone calls continue with "Did you hear?"
And then it all changed

Floors roiling from explosive impact, flashes of flame,
  smoke curling down the hall
Claxon like alarms, directions to evacuate, grab your bag
  and go

But wait—fight or flight? Time to fight!
This is my building, these are my people, time to help
What can be done, have never been here before

A day unlike any other
into the inferno and back again, and again, and again
What would you want if you were the one trapped,
  suffocating?
Time to step up and act

Smoke so thick the taste never leaves,
  gagging for gasps of clean air
Smells of burning jet fuel and human flesh permeate
  the day
Shock on the face of seasoned warriors,
    how did this happen?
Cries for help, from everywhere
Pools of calm, within a vortex of confusion
Is this real?
Never had a day like this before, nor since
A day like no other

GUARDING MARJAH'S STRAWBERRY FIELDS

I remember the music, the muse, the beauty, the blues.
I remember the ruby reds and pretty pinks
  from silky fields of battle streets.
Beaten bare by desert heat,
  I walked along the poppy seas.
Everywhere I looked, a sea of red, a sea of pink,
  like in the land of make believe.
I sang the song as I made my way into town that
  peaceful day.
No one else could see how the colors affected me.
That's when I learned to truly see what was before me,
  the strawberry fields.
When it was time to harvest, they reaped the crop.
When they were done the color was gone.
When the color was gone the beauty was gone.
But those strawberry fields live forever inside me.

PATROLLING THE VINEYARDS, BURNING CONTRABAND, AND PISS TUBES

MSG Martin J. Cervantez has deployed to the Gulf War, Kosovo, Iraq, Afghanistan (three times), and Haiti, where he documented the conflicts through the lens of a camera, as well as on canvas and paper. He used photographs he took of soldiers from Iron Company, 3rd Squadron, 2nd (Stryker) Cavalry Regiment in Southeast Afghanistan for this series of paintings.

In *Patrolling the Vineyards* and *Burning Contraband*, soldiers conduct dismounted village-to-village patrols. These missions, led by members of the Afghan National Army, last an average of 12 hours and involve climbing over endless walls, embankments, and ditches while avoiding IED booby traps and often enemy snipers. During a cordon and search of a local farmer's compound on February 28, 2011, soldiers seized a large amount of hashish and destroyed it by incineration.

In *Piss Tubes*, a soldier stands in the wind, rain, and mud while relieving himself into 4-inch piece of PVC pipe on Camp Iron, a Combat Outpost (COP) located near Kandahar. Approximately 100 soldiers occupied the COP, including several females. Two Porta-Johns were used only for solid waste. The primitive conditions are endured when a unit is asked to stand up a new COP and start with very few supplies. MSG Cervantez never heard a soldier complain about the conditions.

Patrolling the Vineyards, pp. 52-53
Burning Contraband, pp. 54-55
Piss Tubes, pp. 56-57

WEIRD WAR

O so proudly I wore
An Army uniform to Vietnam
On a civilian airliner—
Then was told to stow that gear
And wear civilian clothes
When going off military bases
In the war zone—
It was a weird war—
Funny thing, the Viet Cong
Didn't shoot at us
While we were drunk
In civilian clothes—
But they let us know
How unwelcome we were
When we sallied forth
Decked out in military gear

THE BAGHDAD ZOO

"Is the world safer? No. It's not safer in Iraq."
— Hans Blix

An Iraqi northern brown bear mauled a man
on a streetcorner, dragging him down an alley
as shocked onlookers cried for it to stop.
There were tanks rolling their heavy tracks
past the museum and up to the Ministry of Oil.
One gunner watched a lion chase down a horse.
Eaten down to their skeletons, the giraffes
looked prehistoric, unreal, their necks
too fragile, too graceful for the 21st Century.
Dalmatian pelicans and marbled teals
flew over, frightened by the rotorwash
of blackhawk helicopters touching down.
One baboon even escaped from the city limits.
It was found wandering in the desert, confused
by the wind and the sand of the barchan dunes.

INSIDE OF GLASS

Before, the bushes were silent and nestled
outside protected from the shallowed-out
plains. They were smart bushes
— all sapped up for the winter outside the chowhall.

The bushes even looked like home,
a place now as abstract as tears. But I could
touch the wind as it rippled the earth,

the same earth of mothers and sons,
and the wind carried me a long song of pride.
My blood was not the first blood to soak
the bushes that choked my mouth closed.

Enemies are on the outside of nothing but inhumanity.

Before, I spoke to my brothers like looking into glass,
when our stomachs filled up the same.
But in those tired bushes and my issued-out legs
my brothers melted the glass into my nightmares,
this is who they were to be.

IN AFGHANISTAN, HOMESICK

Lover I love you I love you I love you.
Even agreements that place me above you
crumble in certain ways, certain lights, some days,
nights. I have mountains, monotonous relays,
beauty in bullet-point, dust, and shrill terror,
lives I can end by intent or by error...
Nothing as real as the dream that you gave me:
death-trap apartment, the countertop heavy,
sagged, so we can't even open the drawers,
laminate tile rotting up from the floors—
not very different from this place in some ways.
Maybe not different at all, but the thought stays.

KABUL DOLLS

From the moment I set foot in Afghanistan, I was fascinated by its culture, history and stark beauty—the tragically vandalized Buddhist statues of Bamiyan, the isolated and intriguing Korangal valley, the grandeur and devastation of Kabul. This was a country that had successfully fought off many invaders and absorbed many different ethnicities. You could see this in the faces of the people. Some were as fair as northern Europeans; others had features and complexions that resembled the people of India and the Middle East.

I practiced people-watching from a safe remove, usually while cradling an M-16 on a security detail or from the back seat of a Humvee. I was often approached by groups of children or by men who were usually polite. But it was rare to be approached by a woman. They were hiding in plain sight, engulfed in burqas, quiet and unobtrusive, hurrying quickly to their destinations, never alone.

The *burqa*, a hood-like covering, completely engulfed a woman's head and torso, falling to mid-calf length or longer. Even the eyes were covered with a mesh screen. Most were the same shade of bright blue. Sometimes there was the cheeky incongruity of jeans peeking out from underneath, finishing off with brightly polished toenails displayed in open-toed sandals. I guessed toes were not considered an erogenous zone. But it wasn't just the garb that seemed restrictive; everything a woman did seemed to be circumscribed.

At worst, being a woman in Afghanistan could be likened to being a commodity. Sometimes being a commodity meant you were married off at a young age to a man you had never met, forging bonds between families or even settling feuds. Other times it meant that a woman who stepped out of her place or was perceived to be doing so, was persecuted.

As an interrogator deployed there in the wake of the Abu Ghraib scandal in 2004, you could say my hands were tied, but in a good way. Although I did ask about family ties, rarely did I question detainees about the women of the family since you ran the risk of ruining rapport by asking a culturally insensitive question. I was expected to rely on my better instincts to get the information needed. I learned to listen, pick out what was of interest, and steer the conversation in the direction of the goal, which was actionable intelligence.

Sometimes, however, an offhand comment revealed much. In one particular instance, I questioned a detainee about tribal rivalries in his village. He told the story of how one of his relatives had shot and killed a man from the neighboring compound. He had thought the man was trespassing on their land. He was wrong.

To make up for the "mistake," the detainee's family offered up two of their female relatives to appease the dead man's family. So, in effect, two young girls equaled one dead man. I often thought about what it was like for them to leave their family. Did they understand what was happening? Were they married off to the sons of the neighboring family? Did they ever see their family again?

The Afghan-American translators we worked with often spoke of how Afghanistan had changed since the 1970s when they had fled in the wake of the Soviet occupation. They pointed disapprovingly at the traditional *shalwar kameez*—a long tunic over loose pants, that many of the Afghan men wore—saying that when they had left Afghanistan, you would never have seen a man dressed like that except in a rural village.

Old books and pamphlets about Afghanistan seemed to be stuck in a time warp that slammed to a halt in the mid-seventies. The smiling women in those brochures were dressed in miniskirts and attended the university. Where were those women today?

Most military members were either ignorant about these issues or too busy to care. We were sheltered. Bagram's high security profile kept the majority of service members effectively sequestered from any meaningful interactions with the population. You could go for days on post without seeing any Afghans, aside from the few men picked for manual labor.

Bagram provided many of the material comforts of home. We had a Burger King, a fully-equipped gym and a coffee shop that served lattes. But it offered relatively few opportunities to engage with the world outside. It was a military oasis plucked out of Middle America and dropped into Afghanistan.

ELAINE LITTLE | 65

My duties kept me at Bagram a majority of the time, but early in my deployment I managed to get permission to accompany a military convoy to Kabul. While there, I picked up an English-language newspaper featuring an article describing a company that made dolls dressed in authentic costumes native to the various tribes of Afghanistan. The doll company, Kabul Dolls, took great care with the costumes and craftsmanship. These would make lovely gifts for my daughters, I thought.

When I returned to Bagram, I sent an e-mail to the company asking how I could purchase one of these dolls. The business owner responded by asking me to help him sell the dolls at Bagram. This was more than I had bargained for. I worked twelve-hour days. In the few hours off between shifts, we were expected to handle all our personal affairs and sleep. Where would I find time to navigate the endless red tape involved in securing a vendor's license for an Afghan citizen? I mentally composed the brush off e-mail.

But the more I found out about the enterprise, the harder it got to give Kabul Dolls the brush off. I learned that the dolls were a way for Afghan women to earn money. Many were widows who had lost their husbands during the Soviet occupation or during Taliban rule. But regardless of their marital status there was one thing they had in common: all were impoverished.

Keep in mind that there are not many ways for a poor woman in Afghanistan to make money. In general, they are not free to just apply for any job in the newspaper or on the internet, leave their households for a job interview, or work in a predominately-male workplace. Also many have minimal education. Doll-making made sense because it built on skills they already possessed (such as sewing, embroidery and beadwork) plus the business employed a strictly female workforce. This minimized any scent of impropriety.

I talked to the office handling vendor's permits. They said that if the business owner, Qais Mehri, was accepted he would have his own doll display at the PX. But first he had to make a trip to Bagram for an interview. I was to meet him at the gate and escort him. On the day of his arrival I waited expectantly at the gate. He didn't show.

I was disappointed as I walked back to my B-hut. I hadn't been enthused about the project at first, but now I wanted to see it through. Later I discovered he had been turned away because he hadn't been able to get the dolls x-rayed. All items entering Bagram had to be scanned for explosives. When Qais showed up, the x-ray truck had been dispatched elsewhere. He was told to return another day since the truck was tied up for the remainder of the afternoon.

Several days later after the permit was finally issued, Qais set up his table in front of the PX. The dolls were a big hit. I did my part to drum up business by distributing the doll brochures that Qais had printed. He even invited me to visit the women at their workshop in Kabul. But I knew this would never be permitted. Getting permission to go to a military post with other soldiers or a chaplain was one thing. Expecting to get permission to go with civilians to an unregulated part of Kabul was quite another.

My deployment ended soon after that. Unfortunately, as I learned from friends on subsequent deployments, Qais did not sell the dolls for very long after I left. Vendor's licenses had to be renewed at regular intervals, and I was unable to find out whether he was unable to find a sponsor or if he found traveling to Bagram too dangerous as the war escalated and the use of IEDs increased. I hope that somehow, in some way, he is still supporting and keeping this business afloat.

In the intervening years, I have downsized considerably. I still have the dolls. But I've decided to sell them and donate the proceeds to a charity benefitting Afghan women. However, I'll keep one or two as a way to always remember the women in Afghanistan, as well as Qais, a man who found a way in the most difficult of circumstances to provide these women with an opportunity to improve their lives.

FROM CLERKS

We wrote the paperwork
We got the calls
We booked the flights
for the boys
getting the fuck outta dodge.
Boys with sunken cheeks
darkened by the sun,
sand still in their ears.
Boys with only a burden,
beer, and sex on their minds
to soften the edges of
the sins they carried.
Then there were the boys
that left their burdens
in the desert,
draped in flags
to soften the burdens
of the sins we carry.

LIFE LEASE

On leave,
I heard those civilians talk,
about coming close to death,
and finding God, or a new lease on life.
In war, those moments happen ten times a day:
Understaffed hospital, overworked staff,
six thousand miles from home,
supply lines hit,
gloves to be reused,
no lunch, no dinner,
no sleep, no rest,
mortars start falling,
mortars keep falling.
Welcome to Monday.
Souls teeter back and forth,
like a little girl pulling petals from a flower,
Love life, enjoy life. Life is pointless, screw life.
After the hundredth moment at war,
 we throw the flower down,
Burn the bush to the ground.
We scream.
A new lease on life, three times today, already.

REVELATIONS

When we arrived, it was like a parade.
They threw roses and smiles,
dusty kisses and cheap cigarettes.
Hopeful hands reached out from all around us,
surrounding us with enthusiastic welcome.
Tiny bare feet ran alongside our trucks
on scorching sand and broken asphalt.
They begged us
for bottled water and band-aids, pencils and candy,
little things that we empty-handed liberators
somehow forgot to bring.
And still their little thumbs pointed
toward the heavens
as if we had just descended down like angels,
to pull them up from their rags to our riches,
but the only treasure we brought
was brass and lead.
We were like little boys with handfuls of firecrackers,
running through the streets like we owned the place.
But this was no Independence Day,
and there were no angels in sight.
We stopped mistaking our helmets for halos
the moment things started to explode.
Our wings were traumatically amputated
by shards of ballistically manufactured shrapnel,
shredding our feathers to expose
the bleeding demons beneath.

We were collapsing like buildings,
cracking like concrete,
as childhood myths and storybook fairytales
started to crumble at our feet.
We discovered that sometimes there are no heroes,
and war stories have no happy endings.

Some of us stopped believing in saviors then.

Bayjī
Tigris
it
rao
har

Ba'qūbah

amādī

BAGHDĀD

Babylon
Al Hillah

THE STICK SOLDIERS

To soldiers, I hope the war is fine.
— Girl Scout Troop 472

The children have colored the cards,
dated from December,
with Christmas trees, piles of presents,
snowmen smiling, waving. Sara wants
a doll. Evan, a dog. Kyle promises
to pray for us.

Outside the hooch, we open mail,
hundreds of letters
from youth groups, scout troops,
classes of school children.

Kearns wants to write back,
ask for pictures
of older sisters.

We tape our favorites to the door.
In blue crayon, a stick-figure soldier poses
as he's about to toss
a black ball,
fuse burning,
at three other stick figures,
red cloth wrapped over faces,
Iraki written
across stick chests.

In Jalula, the children draw us pictures, too.

In white chalk, on concrete walls,
a box-shaped Humvee with two antennae
rising like balloons from the hatch.
A stick-soldier holds a machine-gun;
he waves at us,
us, in the Humvees.

Further down the wall, a stick-man holds
an RPG
aimed toward the Humvee,
the waving soldier's head—
what the children want for Christmas,
or what they just want.

HINDSIGHT

A ragged child stares at passing soldiers
Band of babies, brown skin brown dust and dirty feet
Passing soldiers mingle at the market
marking danger with pointed weapons

A ragged child enamored with foreign soldiers
Mister Mister!
Smiling I reply, No, Misses

What a hero I am here in your market, in your country
I'll give you candy and a stuffed rabbit
I won't point my weapon at you
That makes us friends, right?

A ragged child stares at passing soldiers,
passing soldiers, passing soldiers,
for ten goddamn years!

A ragged child grows up watching passing soldiers,
arrogant and self-righteous, disrespectful,
not understanding, not knowing...

The ragged child becomes a woman, a man,
has no memories that don't include passing soldiers,
invaders, occupiers

THE SEDUCTION OF THANATOS

We spoke in glaring whisper. No ums
or ahs but a plethora of fucks
that rose like prayer into an evening
sky licked wet by tracers

We watched the bats flap in their wicked evening
orgies, while the sparrows slept—
nestled on delicate eggs in the cracks
of the stripped Iraqi barracks

Always to the west of us, the lovers
star teased a blush of departed day before
the virgin moon showed the angles of her face.
The virgin winked at us, flirted
and fled—until the night we tasted blood,
we shivered in anticipation.

SALT

The first time I wore my uniform it did not fit
seams still untested
fibers fell which way they pleased
filled out with naivety and a new sense of association
you must wear it right to make it fit
wear it and sweat
until your salt saturates its fibers
while fear flows from every pore
wear it and weep
rasp and gasp for breath
remain in it after the child in you flees
bleed in it, maybe die in it
in which case your cold corpse will fit so snuggly
wear it right and it will fit
salt of your body
blood of your sisters and brothers
may it never fit another

UNTITLED

at dusk
we would leave
our division base camp

walk the red dust road
for a thousand meters

before suddenly turning
into the elephant grass

plunging through a green sea
that sometimes slit the skin

undercover of darkness now
we would move again

to our final overnight ambush site
cloaking ourselves in silence

waiting for an enemy
invisible as the wind

WALT NYGARD, *Marine In Elephant Grass* | 83

SOLDIERS, WITH ANCIENT WEAPONS, ATTACKING

Across a vast expanse
of semi-conscious recollection
of muddied, bloodied fields
where only rats and lice live
are men,
forever advancing,
forever holding useless barricades,
lines of trench, scarred ridges—
the futile and sacred
real estate of war.

East of the I Corps Bridge
on the north side of the river
sprawled a slum called Iron City
a weepin', swampy unforgiver.
Nearby a busted pillbox,
a relic from the French,
squat sideways where it landed
astride an ancient trench.

New guys checked out gas masks
and fam-fired M-16s
with ghosts of Legion soldiers
blown to smithereens.
Rusted wire, rotting sandbags
sunk into the mud,
rats and lice and leeches
came looking for some blood.

A young Marine stared 'cross the river.
The sun exploded, then it died.
The sky went green, then blue, then black.
Flares and tracers soon replied.
He peered through night alive with death,
where the river's a serpent's crawl
the land's aswarm and the jungle teems
and the war devours them all.

I USED TO BELIEVE

I used to believe I would run up the mountain with heavy boots, tired arms, sweat-filled brow, and rifle in hand;

I would hand signal my point man to check behind the wall, and he would respond with the exact location of our target;

I would call in an air-strike on the enemy tanks as they crested the hill;

I used to believe I would yell to my men, "COME ON" as we overthrew the mountain, bringing hell with us, later returning with glory in our hearts.

But now I know.

I know I ran up that anonymous mountainside with heavy mud-caked boots, tired arms, drenched face, and rifle in hand;

I hastily signaled to my point man to check out the wall up ahead;

I had air support above, somewhere, while we chased down a white pick-up truck, about 300 yards away, a mortar tube supposedly planted in the bed;

I yelled "COME ON" to my men as we ran

the opposite direction,

back down the mountainside,

and let that white truck go.

And now I know.

A few weeks later in that same area, Sgt Allen was turned to jelly by an IED that flipped his LAV.

I believe that white truck is still driving.

AL-'ALIM

Allah, the All-Knowing

Acres of wheat, sunflowers,
and a tractor peppered with rust
separate our guard tower

from the minaret.
In the first moments of dawn,
the *muezzin* begins

the *azan*. His call to prayer
rings from the balcony above Balad
and carries across the field

over two men, so close,
we see fog gather above their bodies.
These *shuhada*—

martyrs to be buried
in their bloodstained clothes—
were destined to miss,

because this time
Allah chose us
and our bullets.

CROSSING A MINEFIELD

In one helmet of fear—the blast—
you imagine your feet, arms
your legs, but mostly your balls
blown off & your head.
You envision walking... crawling... passing
that last inch of life into death.

Life ending could go like this:
slivers of shrapnel rip into the body—
pain cuts gash in the imagination deep—
the hollow of the brain shuts down—forever.
Why is fear a weeping eye, a crown of thorns?
Why do I feel someone
is driving nails through my feet?

In the battlefield of hate
what will happen to the bodies of men?
What will claim their miserable souls?
What high moment will stop them
from shrieking as they cross an open field?
What heart will find them worthy of love?

Within men's passion of triggers—
the flashback which happens faster & slower &
 back again—
Beneath the scar—
the deep incision in the souls of men—
is there a moment & a person
that may make all the difference to them?

REMEMBER

I don't remember their names.

Carpenter, Cross, Shields, Garcia, Polk, Colindres, Rivera, Scott, Davis, Feliciano, Mawson, Hesser, Bonner, Merola, Stati, Cordova

I don't remember their names.

Messam, Lopez, Graham, Irizarri, Johnson, Jonston, Preston, Stockton, Rogers, Rogelio, Rigel, Nigel, Victor, Tullock, Travis, Jarvis

I don't remember their names.

Jones, Orozco, Viviero, Smith, Smith, Smith, Smith, Smith.

I don't remember their names.

March 2002, Cave searching mission, Afghanistan. Operation Anaconda. Killed in Action.

I don't remember their names.

June 2002, ARCENT Headquarters. Camp Doha, Kuwait. Raped at the latrines at midnight by four other service members.

I don't remember their names.

August 2003, Pre-deployment training. Okinawa, Japan. Drug overdose. Dishonorable discharge.

I don't remember their names.

December 2003. Web belt and Temazepam. Hanging in barracks, Main side. Camp Pendleton California. Dead on arrival.

I don't remember their names.
But somebody should.

BEYOND ZERO: 1914-1918

These prints accompanied the premiere of *Beyond Zero: 1914–1918*, a new work for quartet with film on April 6, 2014 at Cal Performances, Hertz Hall, Cal Berkeley. This project combined the collaborative efforts of Kronos Quartet, Alexsandra Vrebalov (composer), Bill Morrisson (filmmaker), David Harrington and Drew Cameron (creative consultants), Janet Cowperthwaite (producer) and Kronos Performing Arts Association (production management). Thanks to the hosts and artists who contributed their insight and images including the Magnes Museum for hosing the symposium and workshop and the MacArthur Museum of Arkansas Military History for allowing me access to some incredible portraits of WWI soldiers to use in this project.

GAME OVER

Maybe it was our aimless idealism,
or misguided patriotism.
Maybe we just didn't understand what we were
getting ourselves into.
Driven by the desire to serve some greater good,
or follow in the footsteps of our fathers.
Some of us needed money for college,
or wanted to see the world in a new way.
But we couldn't find any other way to pay for it.
We wanted to prove that we were men,
So we took the oath,
then traded our Nintendo's for M-16's.
We started out pushing buttons,
ended up pulling triggers,
and it was like no game we had ever played before.
When we first tasted war, we became addicts.
Got hooked on bullet crack and artillery blast.
Every day one of us would overdose on sniper shot or IED.
Some of us never got sober, relapsed on 3 or 4 tours.
Wishing we could just go back
to playing video games or Cops & Robbers again.
But there was no reset button this time.
No getting back up after counting to ten.
Our eyes were sewn open,
unable to look away from horror stories
unfolding in front of us.

We came home incomplete,
with no words to describe
how our hearts are now beating us black and blue
for some of the things we had to see and do.
We all got what we asked for, in one way or another.
But maybe we bit off more than we could chew
and we're still struggling to swallow it all.
We're still choking on the truth that we were lied to.
And even though they said our war was over,
it sure doesn't seem that way.
But hopefully,
one of these days,
we'll finally make it back home.

REUNION

Had it been five years ago,
  the first chords of Family Tradition
would have had me out on the wood floor,
  lushing over slurred words and vivid truths
I knew as a war veteran just returned home.

Old dingy bars kept us calm
  for so long until emotions set in. The thought
of not drinking with James or Sanchez,
  Procopio or Pedila or the other souls,
God rest their names, that fled the ruins
  of Ramadi without justice at their grave
or with a flag, folded neatly and rested into
  their family's hands while rifles crack
diligently over war hero's graves.

We left shots on the bar,
  waiting for their ghosts to sip.

As nights went on, we remembered to be men,
  finding a sulky love who would love us
no different than death, or hold deferent wishes
  of lust- the quick fucks that can only compare
to mortar blasts and screams of children
  twelve years old, before brain matter fell into
the gloved hands of medics, before zipping closed
  their black sarcophagi previously rank with immortality –

Hajii- dead to his own kind, never to see
  his mother again, while mine throws back
shots to indulge, trying to ease a war from
  crying with death.

DROPLETS

If I inked three tears high on my cheek
Would you believe I could cry?
If I drew two wings upon my back
Would that mean I could fly?

With a question mark on my forehead
Would you still keep asking why?
If I write "Forgive Me" across my chest
Does that mean I'll never die?

When I take the names of KIAs
and scroll them on my arm
Its length will make you stop and stare
and cause you some alarm.

Youngsters will clutch their mother's thighs,
but I mean them no harm.
I consider it a talisman,
a twisted good luck charm.

So when I take the artist's chair
and lie down on my back
The tattoo gun will start buzzing
sounding the attack.

The needles spit out many colors,
reds and greens and black
But rip the flesh right off my bones,
it still won't bring them back.

WHEN I SAY I AM A SOLDIER

Inspired by Michael Anthony's "American Soldier"

When I say I am a soldier
I am not a bringer of peace
But a destroyer of families and nations

When I say I am a soldier
I am not supported by my country
But brought down by its negligence

When I say I am a soldier
I am not proud or strong
But betrayed and broken

When I say I am a soldier
I am not the image you see on the posters
I am the shutout you do not want to see

When I say I am a soldier
I am not proud of the medals that hang from my chest
But honored by the memories of fallen friends and the scars
Earned in the protection of others on the line

When I say I am a soldier
I am no longer fighting for a country's greed
But fighting myself to forgive and forget

GLORY AND ISOLATION

The gulf between us...
is bigger than the oceans and seas
that took me to the shores of war.
We live in the same country
but I am banned from your reality.
Somehow you relate more than I do
to the generals and commanders of war.
They hold their heads high
during dog and pony shows.
Gleaming sophistication of precise equipment,
yet, in the shit, they come up short,
for both us and the liberated civilians.
But here everything is a spectacle to be enjoyed
and a reserved seriousness
for the authority of their power.
They led me to war
but they seem to have you
completely wrapped up in a military utopia.
I wish you could understand,
even for a minute,
what these spectacles mean to another people,
I relate to more than I do to you.

Don't you care enough to know why?

SONG OF NAPALM

For my wife

After the storm, after the rain stopped pounding,
We stood in the doorway watching horses
Walk off lazily across the pasture's hill.
We stared through the black screen,
Our vision altered by the distance
So I thought I saw a mist
Kicked up around their hooves when they faded
Like cut-out horses
Away from us.
The grass was never more blue in that light, more
Scarlet; beyond the pasture
Trees scraped their voices into the wind, branches
Crisscrossed the sky like barbed wire
But you said they were only branches.

Okay. The storm stopped pounding.
I am trying to say this straight: for once
I was sane enough to pause and breathe
Outside my wild plans and after the hard rain
I turned my back on the old curses. I believed
They swung finally away from me ...

But still the branches are wire
And thunder is the pounding mortar,
Still I close my eyes and see the girl
Running from her village, napalm

Stuck to her dress like jelly,
Her hands reaching for the no one
Who waits in waves of heat before her.

So I can keep on living,
So I can stay here beside you,
I try to imagine she runs down the road and wings
Beat inside her until she rises
Above the stinking jungle and her pain
Eases, and your pain, and mine.

But the lie swings back again.
The lie works only as long as it takes to speak
And the girl runs only as far
As the napalm allows
Until her burning tendons and crackling
Muscles draw her up
into that final position

Burning bodies so perfectly assume. Nothing
Can change that; she is burned behind my eyes
And not your good love and not the rain-swept air
And not the jungle green
Pasture unfolding before us can deny it.

OPERATION RESTORE HOPE

It's not everyday that you see a
Somali woman with her children
in a New Jersey pizzeria.
She said that people often ask her
if she's from the Islands,
her eyes as bright as her smile,
Somalia shining through her chocolate skin,
hair no longer laden with the parch winds
of Africa but with the winds of a blow dryer.
No textiles to robe her from her head to her feet,
the Gap version of her former self.

Assimilation:
for becoming American she pays through her children,
their tongue no longer her own.
"Iska Waran," I said.
"Ah! You know Somali?"
A confused look comes over her face.
She searches for the answer
in my fair skin and freckled arms,
as my son places a quarter in
the gumball machine
behind me.

"Mogadishu," I said.
Her jaw drops
as my son's gumball falls to the dirty floor.
"Oh no!" she says, "May he have another quarter?"

"Sure, thank you so much," I say.
She slips a quarter into my son's hand.

Suddenly, all of the Somali children screamed in my
head as I watched her son and daughter:

"WATER!

WATER!

WATER!

FUCK YOU AMEDIKKA!
AMEDIKKA!
FOKE YOU!"

The guilt of hell fire raining on her people
rips through my being
transporting me back to convoying at full speed
heart pounding
blood racing
past begging children
under the facade
of a humanitarian mission.

My son pops the bright green gumball into his mouth.

Peace had been restored
to my unknowing American boy.

I'M WONDERING

I'm at Guantanamo Bay
Deployment: an individual augmentee for OEF
Operation Enduring Freedom
to JTF – Joint Task Force Guantanamo

I'm wondering
Why is Guantanamo called
"The Pearl of the Antilles"
Why are prisoners called "detainees"
and why don't they have names
Unless you consider "ISN Number 237" a name

Interrogations:
getting information from the prisoners of Guantanamo
theoretically could prove valuable, but

I'm wondering
Does that theory hold up
when the vast majority of prisoners
have never been charged with a crime
let alone some type of terrorist offense

More than 500 have been released
They weren't terrorists

When prisoners are taken to interrogation
why are they told they have a "reservation"
A reservation for
"enhanced interrogation techniques"

or for torture and abuse

And I'm wondering about the cellblocks
Why is the official term "MSU"
Maximum Security Unit
Why not "solitary confinement"

I'm wondering about hunger strikes
Why does the military refer to them as
"long-term non-religious fasts"

The prison camp psychiatric team makes me wonder
Why is it called the "Biscuit"
the Behavioral Science & Consultation Team
BSCT

The prisoners get issued items: a blanket,
toilet paper, a Styrofoam cup, a bed roll,
and a Quran.
I'm wondering
Why are these things called "comfort items"
Do they really make the prisoners
feel more "comfortable"

There are many kinds of people at Guantanamo
(*not* including the prison population
which has represented more than 40 nations)
I'm talking about members of the
Joint Task Force Guantanamo:
the FBI, the CIA, DIA (Defense Intelligence Agency)
and other federal intelligence

and counter-intelligence personnel
And I'm wondering
Why are these groups referred to as "OGA"
Other Government Agencies

Prisoners have been at Guantanamo for over a decade
Almost all are being held indefinitely without charge
For those few who have been charged,
the war crimes tribunals at Guantanamo
are just now getting started
And I'm wondering: Why is the location of the
war crimes tribunal court called "Camp Justice"

The official motto of JTF Guantanamo
is "Honor Bound, To Defend Freedom"
West Point taught me:
Duty
Honor
Country
Don't these three words mean to defend the
Constitution and uphold the Geneva Conventions
Didn't the President, the Commander-in-Chief
sign an executive order to close Guantanamo

So I'm wondering: Why is it still open?

SORRY FOR NOT BEING SORRY

Dear Crab-boy and Street-children Gang of the
Camp Eggers Checkpoint, Kabul, Afghanistan:

I applaud you and your friends' industrious attempts
to trade AAFES POGs for U.S. dollars but I was
angry when you and your group mobbed me. I was
only trying to make my way towards my armored
vehicle. En route, one of your members pursued me.
When I tried to open my door and get inside, he
stood in my way, prevented me from closing my door,
and then shouted, "It's my country, It's my country."
At that moment, he was not a child. He was the
enemy, keeping me in the open and thus causing
me great anxiety, which welled up like steam in a
teakettle.

To remove the threat of that tiny little enemy, I
kicked your friend, square in the chest, and for that I
am not sorry. However, I am sorry that I cannot feel
sorry. I hope you will never have to be who I was,
and hurt those desperate for survival and love.

My main regret in all this is the circumstance in
which I found myself. I was a man carrying a gun in
your war torn country that lost sight of mercy. All
I was worried about was one of your countrymen
pointing a gun at me or readying a bomb rigged to
explode at my feet. In my eyes, and to this day, all I

see are potential threats. That day, instead of offering goodwill and humanity to a child begging for candy or a POG trade, all I could give was the sole of my boot. I kicked your little friend in his chest. My seven pound foot crushed his little ribs into his rapidly beating and innocent heart.

Over the last year, I have spent a lot of time reflecting at my little desk about our engagement. When I think about regret for the things I have done, I am often taken back to the exchange between your friends and I. I was an angry young man with a gun strapped to my hip. The weight of it was like a hot burning coal, stuck deep in my pocket, buried under debris, which made it hard to remove. It burned me. In turn, I burned you. I guess it was a subconscious effort to take the heat off me by projecting it out and away from myself.

I hope you can forgive me. Not just for my sake, but for your own. Carrying hatred from your past will only destroy you and those you love. Trust me, I know. In this life, what we carry is what we give.

Semper Fidelis and Godspeed,

Alex

THE NOISE REMAINS

Sunlight reflecting off a stainless steel coffee maker
forms a howitzer round
on the refreshments table at the university job fair,
where I try to convince myself
a nine to five is the purpose of my life.

The zip-tie holding up the muffler on my neighbor's
  Suburban
reminds me of blaze orange detainees,
their wrists bound, heads sandbagged
and slouched in shameful prayer.

A Lysol-clean classroom tastes like the Dental Corps tent
where I "manned up" for a root canal
because they don't give Novocain in the combat zone,
not to soldiers, not to kids.

Can you hear the chalice drums,
louder than cannon fire,
the beat of five hundred thousand dead hearts?

I confess.
I enlisted for an honorable crime
to erase a felony from my record.
I pointed my rifle like a rich man's cane
to direct Iraqi teens mixing concrete,
slaves constructing a perimeter
to block mortars shot from the dusty fields
of their fathers.
Teeth fell from the mouths of those curious boys

who got in the way, who reached out
to touch Kevlar skin, coarse hands–
and the same thing now happens to me
when I open my mouth to speak,
standing naked and bone thin
before an audience of ghosts
in the worst dream of all

this is who I am,
who we've become.

Jessica Lynch
is not my name.
David Petraeus spits in my face
when he writes "How We Won in Iraq."
Did George W. Bush
really happen?

My story is my own,
and it ends in loss.
Like a gunshot in a basement,
the remains of a thousand explosions shout
(buzz through my skull)
Iraq
drowning out
the moral of these words.

And what remains is a burning whisper:
to enjoy myself
is to forget our routines
are held together
by fragile agreements
tested and broken in war.

SERVICE AND WHAT REMAINS

I've talked to many families who know someone who has served. Often, they say the person "doesn't want to talk about it." I experienced this myself. I bottled up all my pain, drank excessively, and broke down while huddled in the corner of the room. As I began learning different methods to express myself, the passion I poured into these works healed me. My art has become more about why we sacrifice our lives rather than the pain endured while in service.

Kurdish Man is from the "We Are One" portrait series, a project that began as a way to look at life and test my technical ability. It grew out of my interest in the way other people live. Like many, I served our country to improve my "quality of life." However, coming home, battling PTSD and depression, and reintegrating into our culture has forced me to reconsider what is truly important. Being around other veterans and being given a way to express all those emotions released a wave of ununified art, including *WPP7* and *Lead*.

"The Bondo Project" focuses on the perceived relationships between success and material belongings and the values and norms passed between generations. I used automotive repair epoxy and plywood, the same tools people use to make repairs, to create *Ronin* and *To Serve and Reflect*. The work aims to understand the psyche of those struggling to maintain their perception of success and the lengths individuals will go to to repair and maintain that dream.

Kurdish Man, p. 119
Lead & WPP7, p. 120 & p. 121
Ronin & To Serve and Reflect, p. 122 & p. 123

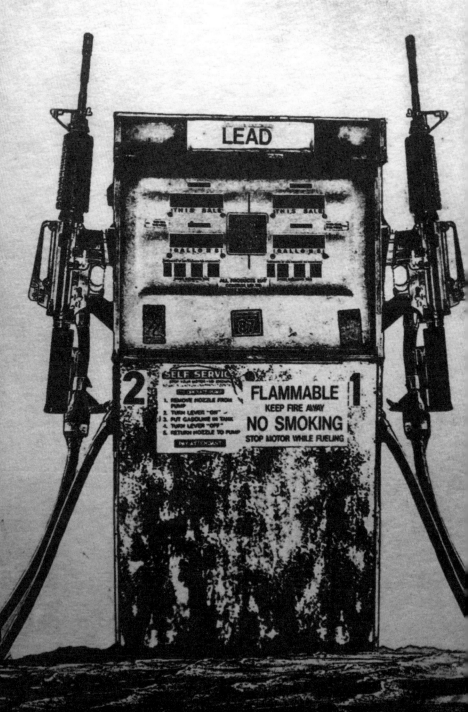

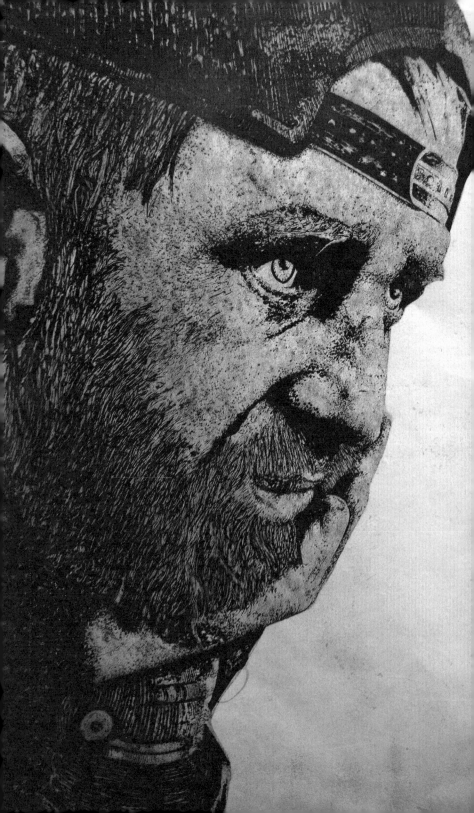

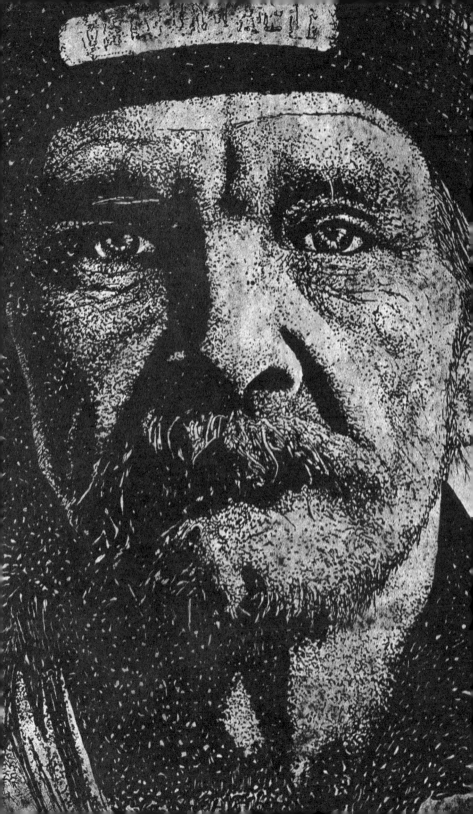

WITHOUT A TRACE

My memory is becoming
a remnant of rough cloth,
worn and tattered.
Dust and mold grow
thick in the aging mesh.

Thoughts, now less than souvenirs,
fade and fall, silent
among the rubble at my feet.
Bitter fondness creeps over my soul
with steadfast unkindness,
joining the weary guilt
already locked around my heart.

SECTION 60

Arlington National Cemetery

Headstones mark their advance
where they hold the line

east of Eisenhower Drive,
and each is decorated:

hearts, crosses, and stars,
flowers and pinwheels,

bottle caps of favorite beers.
Dog tags chime against marble

and heaps of dark soil settle
over the youngest plots.

A groundskeeper follows orders
and starts digging in,

while visitors inspect the ranks.
Each fresh-faced stone,

rigid and upright,
holds its place at attention,

as families mark time
by the procession of seasons since.

HAPPY BIRTHDAY

June 9th 2000: The Devils win the Stanley Cup!
Garett's carrying a keg above his head!
This place is CRAZY!!!

June 9th 2001: Twenty one! Limo! Downtown Denver!
Avs beat the Devils in the Stanley Cup!!!
Our clique, The Hungry Strangers are born!!!

June 9th 2002: Wow this year has been crazy...
Joined the Army, 9/11, boot camp...
Now we're in Germany drinking
at O'Shea's for the birthday!

June 9th 2003: Kosovo... This place is hell!
Cold in the winter, hot in the summer.
What the FUCK are we doing here?

June 9th 2004: Damn, I wish we were back
in Kosovo, Iraq sucks! 130 degrees outside today!
And guess what, getting shot at on the birthday?!

June 9th 2005: Been out of the military for nine days.
Should I be happy? My grandmother's in the hospital
and is going to die. Why don't I feel anything?
Why don't I care?

June 9th 2006: One more drink to help me forget
One more pill to kill the pain
Fighting for the numbness that I hate so much

June 9th 2007: I'm told it gets better, if you let it
That it gets better if you try
I try to put the pieces together, remake sense of it

June 9th 2024: The bombs still echo
While the candles still burn
Distant memories serve as lessons for today
To another year hopefully not pissed away

BARBATO

In Memory of John Joseph Barbato

It's a long, Long Island,
when trading breaths for
euphoria, and
slipping sideways, towards
inward.

I can feel John, in
the heavy thick
molecules, settling
dusts of many, and
the collapsing lift of all forms.

I can smell John,
in the absence of summer,
leaves carrying his scent, in
the early approach to the,
fall.

I can see John, sick and struggling,
flashing
cloudy eyes,
from a storm swept brain.

I can hear John,
ghosts escaping in,
mangled muffled battles. He
was listening for
colors, in
a black and white world.

It's a long, Long Island,
when trading breaths for
euphoria, and
slipping sideways,
towards inward.

JOSHUA CASTEEL

RIP 1979-2012

Iowa blond
Midwest height
Seeking self
Seeking others

Respectful smile
Humble eyes
Deep impressions
Nuanced rhetoric

Firm body
Bookish glasses
Tranquil fighter
Conscientious witness

Sport coat
Sox cap
Desert winds
Hidden assassin

Loving embrace
Stern vision
Brilliant faith
Transcend conflict

Just rest
Hopeful sage
Love one's
Enemies

130 | AARON HUGHES, *Burn Pit*

BURN PIT CANCER

Burn chemical, biological, radioactive, nuclear waste
Burn intelligence
Burn batteries
Burn emotion
Burn garbage
Burn instinct
Burn concrete
Burn meaning
Burn tires
Burn sadness
Burn shit
Burn hope
Burn dirt
Burn lust
Burn acid
Burn faith
Burn flesh
Burn love
Burn pits
Burn
Burn everything

Get inflammation
Get humiliated
Get discharged
Get disgraced
Get poisoned
Get cancer
Get growths
Get lost
Get sick
Get sad
Get dead
Get
Get everything

TO THE IRISH AMERICANS WHO FOUGHT THE VIETNAM WAR

In the moans of the dying Viet Cong,
From my Gran Da's tales, the Banshee.
In the calmness of prisoners shot for spite
The brave James Connolly.
In the hit and run of those we fought,
The Flying Columns of the IRA.
In Tet, so unmistakably,
That fateful Easter Day.
In leaflets found in farmer's huts,
The proclamation of Pearse.
In all the senseless acts of racist hate,
I felt the growing fears.
In the murder of unarmed peasants,
With our modern technology,
We became the hated Black and Tan,
And we shamed our ancestry.

FINE

I am fine, how are you?
Fine you say. We smile and move on
We are both lying, neither one of us is fine
Yet everyone we meet, we play this false game

Why do we do this?
Are we too busy to say what is really going on?
Are we lying to ourselves as well?
Are we Avoiding the truth?

Things are not "fine"
Unless environmental degradation is fine
Unless being a wage and debt slave is fine
Unless endless war is fine
Unless broken people are fine
Unless food insecurity is fine
Unless fear instead of love is fine
Unless, Unless, Unless...

There is a sense of disquietude in the land
and it is also inside of us
We all feel it
It seems too large to deal with
so we say it is fine
As a people we are like the fabled ostrich
sticking our heads in the sand
or the monkeys who can't see, hear, or speak of this evil
We are meant to be so,
                    so much,
                            so much more.

And yet we continue to live in a world of lies
And that is "fine"

SILENT WARRIORS

Last summer I was at a gathering for an elderly couple relocating to assisted living. It was a cheerful occasion and I did not know anyone other than my partner and her immediate family. Instead of fraternizing, I sat on a picnic table in the driveway and drank a beer. After some time, my partner pointed out the couple hosting the party. The woman appeared much older than mid-eighties. She wandered around the neatly mowed yard mingling cautiously, her head tilting from side to side. The man stood tall with a baseball hat perched intently on his head. His silver gray hair was cropped neatly and his cleanly shaven face glowed in the afternoon sun.

As he stood with his hands on his hips, I noticed something. Fastened to the middle of his belt were five men hoisting a long slender flagpole with the red white and blue defiantly waving. Illustrated on his belt buckle was the symbolic image from Iwo Jima.

Over the years, I have battled with my military experiences. Many of the things I did still haunt me. Do veterans of previous generations feel this? Do they regret their actions? I had never asked. As the man sauntered around the party for the next hour I felt obligated to ask him about his belt. When I mentioned it to others they said, "He never talks about it."

I gathered my nerves as I stood up. My confidence grew as I stepped forward. Approaching the veteran from behind, I asked, "Excuse me sir, but can you show me where the bathroom is?"

He turned to face me and replied, "Yes, of course."

Leading me inside, he pointed to the right and said, "There it is."

As he met my gaze, his clear blue eyes pierced my thin layer of confidence and I nearly backed down. Without hesitation, my eyes quickly darted to his belt buckle and I asked, "Were you at Iwo Jima?"

"No, Okinawa."

Feeling nervous, I jutted, "I am an Iraq war veteran and wanted to thank you for your service."

I am not sure what I expected him to say. Were we to embrace like old pals or tell stories over a glass of scotch? No. Instead, I received a gentle pat on the shoulder. "Thanks," he responded, drifting back to his party.

Was that it? Maybe this guy did not give a shit that I am a veteran or that I have been to war or that I have driven all-night missions to Fallujah. Maybe he did not care that I walked miles through fields flooded with shit. Perhaps he did not think twice about it, because he did the same thing, at a different time in a different country.

Perhaps he did not want to confront the truth that there is a whole new generation of young men and women dealing with the same old problem. This man knew what PTSD is and could not sit through another war story. He had enough blood on his own hands.

Back outside, the warm breeze hit my skin and I felt the trees sway with my eyes closed. This was not what either of us fought for. Why do we continue to send our youth off to die, returning dismembered or mentally damaged? The cycle must stop. Our society must stop creating generations of silent warriors.

NEW YEAR'S EVE

I visited with old friends today: the friends I made
  in uniform, the friends I made in Baghdad
We caught up, told stories, shook hands,
  and slapped backs
We laughed about our weight gains and about our
  graying and thinning hair
We smiled and hugged and were genuinely interested
  in how each other had been
It was a beautiful December day in Oklahoma
  and we were together again
The sun was out, the air was crisp,
  and birds flew thick overhead
But just as soon as we visited, the funeral was over,
  and it was time to bury our dead

Killman was gone and here we were laughing
  and joking at his funeral
Slowly, the guilt overtook me and it stayed with me
  all day
I realized we hadn't spoken since we returned from Iraq
It made my stomach turn
And I wished I could turn back the clock
He was just 29 years old: a son, brother, husband, friend
He was funny, tough, reliable and reserved
And he was always smiling

I was shocked when I heard the news, we all were
He was married one month before;
  he just celebrated his birthday
And it was two days before Christmas

I want to take comfort in the proverb
  "still waters run deep"
because that would absolve my guilt
I didn't know it was coming
but the truth is: I didn't know because I didn't care

I didn't care enough to stay in touch
I didn't care enough to ask how he'd been
I didn't care to ask what he was doing
And it feels like my greatest sin
Seven years after returning from Iraq,
  my friend is the latest casualty of war
But there were no 21 gun salutes
And like so many others, his death won't be counted
The Pentagon won't care
He's just another veteran who committed suicide
And I'm another buddy who wasn't there

Maybe things will change
Maybe I'll feel differently, in a month or in a year
But right now I still feel guilty,
  because my friend is dead
And I've yet to shed a tear

WHAT WAS BAGHDAD LIKE?

A constant feeling of death
comparable to;

being bound tightly   then

continuously dunked

into a shark tank.

REMEMBER TO BREATHE

A subtle transition to what is next and it's time to
  open heavy doors with many locks for which I have
  no keys
I have been reluctant to peek around corners for fear
  of discovering untrustworthy eyes
I've been sweeping the ground with my foot and
  continuously looking down even though the
  landscape is so radiantly beautiful when I look up
This frozen reservoir has had less than its fill of rain
  and snow but far more struggle than its weary dam
  can hold
The moon throws sucker punches at it all night,
  bouncing them off Jupiter and making the sky black
  and blue from contusions while red and purple
  mornings beg for mercy
These barren rocks look like gravestones that are
  ecstatic to have such astonishing views
The sun peeks over the horizon just in time to kick
  everything out of my dreams

Strangers with fiery eyes repeatedly ask me if I'm
  lonely as though they are offended by a man
  drinking alone in public
The look in my eyes tells them that there is little
  room left in my life for smiling, but goddamn it,
  I try
I'm often early for happy hour but always too late to
  let anyone through the back door after hours

There is sympathy and thanks written on yet another
  note from a stranger as my story becomes me
Yes, you may have read about me in the papers and I am
  a veteran
As if I didn't wear it on my sleeve, as if you couldn't tell
  by my complete lack of presence

My jaw clicks with each drag and my shoulders tighten
  to hold my heart up as it bangs a tuning fork off my
  spine to play the perfect pitch of my hands shaking in
  nervousness
When did that butt can begin to overflow? I don't
  remember stepping outside so many times because I'm
  afraid of shooting stars
Joe called me late last night to remind me of that time
  he thinks I saved his life
A cut wire mishap and a strong survival instinct saved
  me
I'm not speaking metaphorically when I say that the
  gunfight I planned on having with myself was
  cancelled on account of the armory being out of
  ammunition
And I've had years of "why am I not surprised" moments
  like when that Frisbee collided with my face after
  reconstructive surgery, or that time I woke up in my
  car breaking through the ice in that cold canyon creek
I don't want to give anyone too much credit for the damage
  report because everyone has their breaking point

On the sidewalk, someone inscribed a subtle message
  within a chalked out heart that read, "Remember to
  Breathe"
Since returning home, it has become increasingly
  more difficult to breathe
So I've been picking up the pieces of my skin that
  finally shattered after a sledgehammer pounded
  these images of war and cavalry into my body
. repeatedly
Snow begins to cover the freshly scattered ashes of a
  dead banjo lying on the ground to remind me of the
  last time I was able to breathe without a prompt
Don't feel anguish for my bleeding eyelids that have
  been seared and stitched shut
I'll be just fine. It just may take forever to put all the
  pieces back together again

ONE YEAR IN A VA WAITING ROOM

A year, a week, a day
At war minutes pass like in a waiting room VA
Seconds low crawl by
Like a dust off high
Pulling tension back on the clock hand

I just can't stand it
This dull alien planet
Florescent tubes flicker
Fully automatic belt fed boredom
I wish the end would just come quicker
Is it *Haji*, *Mujahideen* or me on the trigger?

Like some hospital patient
impatient for some action
Cursing yourself for wishing it
A violent satisfaction

That's the same plastic Christmas Tree we had in Iraq
Why is it still up with this Easter crap?
Holidays slip past without notice
Like a hand waving in front of a thousand yard stare

No major events to divide the missions
Up MSR nowhere to guard some provisions
Rush hour traffic, you're late to clinic
Road rage, roadside bombs, an ambush, I panic

A man with a clipboard takes my blood pressure
Where is the pharmacy?
Can you please define "stressor?"
The VA's a war zone between me and my home
The closer to leaving the longer it takes

IED. Contact.
RPG. Contact.
Mortars. Contact.
VBIED. Contact.
Home

EMOTIONAL CUTTING

When a deployment to Afghanistan is appealing
Something is broken

I want to grow old with somebody
And can't take that from anybody
Do I answer the call of my soul
Or silence it to a low hum buzzing tolerable but there

It is better if I lie to myself
Safer and less harmful
The racket of a guilty conscience is silent to the ignorant
Informed exposure is masochism
The less I know about what I am doing is wrong
The more I can identify with those yellow ribbons
Allowing the noise of a need for belonging
To drown out that nagging conscience
What about after the tour
What happens when I get back
Where there are places of silence and I can hear my
    heart again
I think this is masochism
An emotional cutting of my soul

The awe glowing from those whose eyes once saw
    only my darkness
My self image is no longer based on your palate for
    character
But by my moral tastes

"Big strong gay boy joins the infantry"
The ultimate proof that I am not that kind of fag
Validation
I love to be a queen and I love the world I live in
Would more combat give my voice volume
Give me a platform to meet my goals
Is this the price for a platform
Self-sacrifice in the humble interest of the world
SEE!!! Look what I did for you
Our pain, forever fresh in their eyes
So too will I use my emotional combat wounds

We have access to the American heart where most
 do not
Our scars are loud, they are heard
Wouldn't exposure to and triumph over obstacle
 make one stronger
And the intensity of the obstacle equal to the quality
 and amount gained
This invites as much pain as possible
Expose myself to the hell of life
My life is built on a platform of pain

Something is broken
When pain and death are invited
But they listen
They listen
Listen

TODAY, EVERYDAY

Wake up to realize I was never asleep
Brain pounding against my skull
Shattered combat dreams
Deaths of lost Marines
Sacred body taken from me in rage
Lost identity, lost mobility, lost smile

Twelve meds down to start my day
Therapy, doctors, sessions lined up
Anxiety of the world, stress filled nerves, memory blank
Service companion Shian by my side
To love and guide me through my time

1600: appointments done for the day
Thanks to my reminder calendar, I didn't miss one
Filling my journal to preserve the memories I need
Painting, poetry, fly-fishing award me joy,
  distraction to close my day

Heavy pull of memories, numb feelings shadow my night
Call from my lover eases my agony
Her support fills my heart, turns my head to a brighter path

More meds to take
Exhausted, I pray for sleep
I made it through the day
Not alone, never alone
I'm strong, I'll recover
I'm determined to find me again

Tomorrow, another chance

YELLOW RIBBON

Originally written as a song

Take that yellow ribbon off your car
Take that waving flag off of your door
Don't act like you understand what we've been
  fighting for.
Take that fake concern off of your face
Take that handshake and just move along.
Stop appreciating me for all that I've done wrong!
And take that yellow ribbon off your car.
You tell me I went to liberate
So when I come home you celebrate
But you can't bring back the dead by throwing a
  parade.
You tell me I made my nation proud
I wish you wouldn't say it so damn loud
My boots were on the ground while your head was in
  the clouds
So take that yellow ribbon off your car.
When you tuck your children in at night
Don't tell 'em it's for freedom that we fight
Let them know that there's a war on but don't tell 'em
  their side's right
And take that yellow ribbon off your car.
Don't pay for my meals, don't give me special deals
Unless you want to hear all about the way I feel
Don't make me your hero, just lend me your ear
Oh, and wipe the tears I cry
While I apologize for that goddamn yellow ribbon on
  your car.

HEADSTANDS

And who knew - that our bodies, outstretched -
Inverted asanas
Right side
Up
Headstands
Necks taut
With tight nooses.
His father found him
Hanging
In the garage & lifting him
Down
Held him in his arms -
We are dying.

We are dying.

And who knew - that our bodies,
Silent and still in
This...shavasana -
Corpse pose
With palms up, body
Like dead weight
Against the ground.
Say what you will Petty, Officer
Say what you will Master Sergeant
Say what you will Lieutenant Colonel
Say what you will airman of the year -
 - You raped them - And

Now they're lyin' lifeless on the floor
With palms up, sticky with pills
And wrists bloody breathin' slowing,
Now stopped.
We dyin' man, we dyin'

We are dying for clarity,
And purpose.
We are dying for memories, and flashbacks.
We are dying, to tell
The truth.

And who knew that our bodies swayin' in the wind
Were cautionary tales of war and violence
And who knew that
Suck.
It.
Up. Meant seroquel tinged - heart palpitations,
These cocktailed concoctions,
so if we don't die by our hands
We'll just die in our sleep.
And who knew that one shot one kill meant a bullet
Through your brain or a bullet through your heart,
you're a thousand miles away
In your own backyard.

Bohica 6 Delta Come in, Come in!
Bohica 6 Delta Come in, Come in!

Your moms about to see
Your boots draggin' in the mud

While you're hangin' there lifeless swayin' from above

And a thousand emotions are running through your mind
Still starin' at the aftermath
Staring at his face -
Swollen left side cheek disfigured -
Black body in dress blues squared away,
In a flag draped casket.
Damn.
For what?

Years before,
Your father found you, on Father's Day.
Swollen left side cheek disfigured
Poison strewn all around -
And a bullet through your heart.
Were you with us, when we buried you,
That day?
You're buried deep,
Stuck, back
In the belly of
My memories - Demons -
Swirling around like violent dry heaves -
Nothing.
And I can't seem,
To let,
You go.

I'm still
Just trying
To figure out who to fight.

Ten years later and I'm still,
Just learning...
How to grieve.
Everyday,
I see your lifeless body, before mine
In my deepest meditations you float -
Facing before me.
God damn.
Dixon.
Can you hear me?
I am struggling,
Stepping nearly faithless forward -
Simply trying, not to fall.
But it's still happening...
And
Sergeant.
I am still.
Just trying.
To follow.
Your lead.

PATRICK

He can't get out of his head the horrors he's seen
picking the pieces up and stuffing them back.
Like Lady Macbeth, his hands will never be clean.

But not like her: she murdered a king;
he was only an army medic, but now
he can't get out of his head the violence he's seen.

He only tried to put the pieces back
into broken bodies. It wasn't his fault,
but now he thinks his hands will never be clean.

His unit invaded Fallujah;
urban fighting's as ugly as fighting gets.
He can't get out of his head the havoc he's seen.

He can't sit still or even look at his hands.
He tried to save his friends, the dead and the maimed.
Now he's sure his hands will never be clean.

The army said he was fine and sent him home
with so much blood his hands will never be clean.
The VA gives him pills. Frozen in time,
he can't get out of his head the carnage he's seen.

ALL THE ANSWERS

My son asked me the question,
  Dad, what was it like?
I said, Son, if you've done it once,
  it's just like fallin' off a bike—
No, that's a bad example,
  it's like getting tangled up in a kite.
You know what, Son, go to bed,
  that's enough for tonight.
Now don't you dare judge me,
  just what would you have done?
Tell him how you killed a teenager
  who didn't even have a gun?
Would you tell him of the battles lost,
  the ones he thinks you won?
I can't keep acting like I'm having fun,
  because my memories weigh a ton.
When he woke without his answer,
  he decided to make a fuss
He asked, Dad, what was it like
  when your boots were in the dust?
Slipping, I said when they yelled, *Allah Akbar*,
  we yelled In God We Trust
And when we shot those motherfuckers,
  they bled just like us.

THE GRENADE SERIES

The Grenade Series are photographic images representing the effects of Post-Traumatic Stress Disorder on soldiers and their families. The grenade tries to symbolize this reintegration and was chosen for its three uses which can be mirrored in those who suffer from PTSD. Grenades are used for attack whether in self-defense or offense, to signal help and to provide cover.

Soldiers who suffer from PTSD often become defensive and lash out due to increased anxiety, they may also seek help to find consolation, or in many cases hide their suffering and detach. It is important to understand their struggles and recognize the battles they continue to face when returning home from conflicts and trauma. The grenade in each image is a sculpture created and incorporated into each scene.

This work is an ongoing series which continues to develop. Stories shared, learned and understood progress into photographs which depict events whether therapeutic, uneasy or familiar. These photographs hope to illustrate the many different hardships soldiers and their families face during the reintegration and healing process.

Tea Time, p. 157 (top)
Say Grace, p. 157 (bottom)
Gifts From Dad, pp. 158-159

EVIDENCE

memories colliding inside my head
twisting, churning, into a jumbled mess
clouding and confusing, which goes with which

the faces seem to be the only certainty
face 1, face 2, face 3, on and on they go
no name, no knowledge of who they once were

only the actions of a moment in time
leaving them forever, a mental picture in my mind

they glare through my fogged thoughts
as if to judge, condemn, and sentence me
with each face I can still smell the powder and dust

I can still feel the gentle nudge of
the stock against my shoulder
triggering the shutter of my mind's camera

trapping that face, that moment in time
evidence for the trial inside

ARMOR WITHIN

He failed so miserably, he will never know
He sucked every drop of happiness,
 joy and wonder from my existence
He made me pray every night that
 I would not wake with the next sunrise
Or better yet, I would just disintegrate
 in the mattress I lay on
He made the satisfaction and amusement for living
 vanish into thin air
Like it was only a memory to mourn for

Covered in the depths of depression,
Melancholy and sorrow
Longing for a life I could no longer have
Yearning for the existence I once knew
My dreams, the future that was once possible
But is no longer
The external battle is now within
The armor I once wore cannot protect me from him
He will never surrender trying to destroy me
The depression and despair keeps lurking
I still feel him circling, I am still his prey
But my body armor within will protect me now

WEST POINT

For Jose R

The Hudson, under a coverlet
of fog. A freight train rumbles along
the cliff face below Jose
& I. I hear thumping, thump
thump & I look up
trying to discern from which direction
I hear the helicopter. There is no helicopter
just a train. I tell no one except you
reader. Keep this in confidence
Thump thump. I laughed with Jose
about the steepness of the cliffs &
he said: *The British were never able to capture*
West Point. Grape shot.
Grape shot would've cut them into pieces.
& the Hudson is dark & deep & cold.
Can nearly hear the British.
Ships shattering stink of Redcoats bleeding.
I sigh
because I know what it sounds like
guns going off I mean.
A light rain the cliffs grey the Hudson hidden.

HOME FROM IRAQ, BARKING SPIDER TAVERN

Cleveland, Ohio

Outside on the smoker's patio,
the Army vet shakes my hand
for the twentieth time, yells
about loyalty, country, duty.

Between gulps, he explains his shame
for missing the Storm—
a bum knee, ten-thousand
beers later, and now, another war
to miss. We finish the cans,
throw them at a wall, crack new ones.

The summer sweat sticks to his face
and in his eyes is the horror
of not going, that he'd live
all his life having to say no,
blaming a bum knee,
hitting it hard with a palm
to punish it.

He shakes my hand again,
grabs my shoulder,
and then seems to want to kiss me,
suck out whatever was left
since he'd wanted to taste it so badly.

CADENCE COUNT

Low retention shows the mood,
This culture can be kind of crude,
We've made enemies along the way,
But a New Dawn, starts another day,
   And it won't be long,
      Be long
   Till new 'cruits come marching on
Victory our governmen' wished,
Fifteen thou to reenlist.
Buy yourself that Cadillac,
Throw your homies in the back.
Forget Jody, Sarge claimed once more,
But I think I've heard this spiel before
   And it won't be long,
      Be long
   Till old 'cruits come marching on
Haven't shaved or cut my hair,
Since that bastard sent me there.
Now I read VA backlogs,
and search job catalogs
   And it won't be long
      Be long
   Till all 'cruits come marching on.

WHEN I SAY I AM DISABLED

Inspired by Michael Anthony's "American Soldier"

When I say I am disabled
I am not asking for compassion
I am simply stating my given label.

When I say I am disabled
I wince in pain and embarrassment as it leaves my lips
Because now I struggle to finish tasks
That came so easily just a few short years ago.

When I say I am disabled
I am not asking for help
I only want to let you know my possible shortcomings,
That my exterior does not match my interior.

When I say I am disabled
I am not asking for your pity
I am only expressing to you
that I am different than I once was.

When I say I am disabled
I want you to know the man I used to be:
Strong, quick-witted, funny and intelligent.

If your mind sees that man in your imagination
Please tell him I miss him.

ASSESSING THE DAMAGE

I was diagnosed with Post Traumatic Stress Disorder
In August 2007 by Doctor Howard Cohen.
However, my mother stated I had PTSD
The moment she saw me when I returned to Germany.
Or was it over the phone when I was still in Iraq?
It does not matter when she knew.
She just knew.
It is always easier to talk about mothers.
They are intuitive, loving and comforting.

My mother made the call to go and meet
The service officer in Philadelphia.
My best friend Melly, flew in from Indiana
To try and explain what was happening
To my mind, my body, who I used to be
And why *she* did not exist anymore,
Because she was living it too.
And my father was the muscle,
I was going to listen.
After the initial week of evaluations
And paperwork in Philly
With my mom, Melly and my service officer,
I submitted my disability claim and got my VA card.

Now, it was time to start assessing
The damages of war
And my father was going to take
That ride with me.

Tuesday was Philadelphia Day with Dad.
Every Tuesday for six years,

My dad used every
Sick day
Vacation day
Family hardship day.
He drove me two hours to the Philadelphia VA
For my appointments and then to a suburb
To Dr. Cohen's office
For my "real" psychology appointment.

Then two hours home, during which I usually slept.

I lived in a cabin tucked back in the woods with a guy.
Who was completely indifferent to nearly everything.
This suited me fine,
Because I didn't need anyone getting in the way of my
  self-destruction.

I could see my dad's car ascend
The driveway from our large picture window in our
  dining room
As I scurried around to get ready.
  He hated when I was late.
Some days he would come in and say hello to my dogs.
I liked this, because it allowed me a few extra few
  minutes
To gather my pieces together.
He gave me the strength to walk out that door.
I was hoping today would be the day he would not
  notice
The stench of alcohol emanating from my body.

I hopped in the car.
"You smell like a brewery."
"Yep."

I wanted to tell him: I can't sleep, I can't stop.
I felt so guilty that I lived that way,
That he saw me that way.
But all I said every Tuesday was "yep."

We descended the driveway and
Every time dad would ask,
"Is your boyfriend ever going to fill these potholes?"
As he grunted and groaned as his nice car bumped
And careened through the very large holes.
"Probably not. He works a lot."
I felt ashamed I put my dad in this situation,
And helpless for the first time in my life,
I was no one.
As I sat there, my hands shook,
Looking at the ground, looking out the window,
Angry, sad, not knowing who I was,
But this man, this Marine,
My dad, thought I was someone and showed up every
 Tuesday.
My mom made my dad coffee
In a to-go cup and mine was right next to his in the
 cup holder.
We both loved that woman.

He picked me up at 7:30 am.
The drive was usually pretty good.
We would talk about football,
Art, dogs, dog training, our family, my medications,
PTSD, the VA and veterans.
He had the play-by-play, week-by-week,
Of my therapy and moods
And is one of the only people in the world
Who could handle my rage.

My rage:
The reason my dad drove me to my appointments
Is because my road rage got so bad
That, combined with having no fear of death or speed,
My parents and Doctor Cohen were worried.
Especially when I started running people off the road,
Having flashbacks, scanning for bombs and swerving
To miss bombs.
Once while driving I followed some dude
Back to his house to yell at him about his tailgating
Doctor Cohen and my parents decided
It was a good idea I not drive for awhile.

Being with my dad was like having my own secret
  service.
In the car, I could relax
He is a great defensive driver and never strays into
  other lanes.
We learned together getting mad at other drivers
Does nothing but get us upset.
He would park the car so I was not late for
  appointments
When veterans were blocking and arguing over
  parking spots in the lot.

Once inside the VA, we sat against the wall in the
  waiting rooms
So we could watch the door.
Or we stood in the hallway,
If a veteran was freaking out in the waiting room.

On one particularly bad day,
I was checking out at the front desk in mental health
  services

And a Vietnam veteran was on a cell phone
And I could not hear the receptionist to set up my
  next appointment.
I told him I was having a bad moment and asked
If he could please walk farther down the hall
And continue his conversation.
He started screaming at me.

I lost it and we were screaming back and forth
And I felt a hand on my shoulder
And my dad said we needed to go.
As we quickly walked to the elevators
We saw security running down the hall to scoop that
  guy up.
But not me.
My dad protected me.
Watched my back.

Then we headed down to the pharmacy
To pick up the newest concoction of pills
The shrink had come up with.
It was usually a two hour wait.
Me and my dad would joke around
And listen carefully to the older vets.
On many occasions one or two veterans were discussing
The best way to blow up the pharmacy.
One veteran explained to us how to place
Sticky bombs on the ceiling.
Another one wanted to set up a line of
Claymores and there was always talk
Of driving through the front doors
With a large truck, up-armored vehicle, tank
Or something of that sort

And shooting up the place.

After the VA circus, my dad drove me
To see Doctor Cohen, my real psychologist.
Doc taught me that my mind was a toolbox
And the sledgehammer was not always the best tool.
He helped me understand that there was nothing
Wrong with me, that PTSD,
Post traumatic stress disorder, is my
HUMAN reaction to the extraordinary circumstances
 of *WAR*
Doctor Cohen wrote me a prescription for a service
 dog
Who would give me my freedom back

My father never lied to me,
He tells it like it is
Even when it hurts.
He never broke a promise to me.
He challenged me to "shake it off"
And do my best.
And people may think that is harsh,
But I think it was great.
I'm a survivor, because he raised me that way.
My dad once said, "If you are a failure, I'm a failure."
So I guess that means if I'm a success,
You are too,
Here's to you, DAD.

FAR AWAY BOY

Far away boy, I can still see you.
Uncontrollable crying, inappropriate laughter,
  I can still hear your muted voice.
Take your time, you are resilient,
  learn to protect yourself.
Fall down and get up, incredible sweetness,
  this world is not your fault.
Far away boy, you are here.

A DAY

Wake up
swallow pride
swallow pills
drink coffee
pull head out of ass
lather, rinse, repeat
blow dry
dress myself poorly
eat poorly
waste brain matter on T.V.
regain motor skills
think useless thoughts
do useless things
drag feet on ground
groan, sigh
pick brain up off floor
pick self up off floor
say "Self, fucking do something"
sleep
repeat, repeat, repeat

LAST CALL

War has a way of touching you at night
It waits until you are just asleep,
Comes to mind and
Reminds you of death,
Kisses your swollen body
With desire and fear
Asking softly,
"Do you remember?"

You weep trying to shutter away
The memories
That haunt you
    The fateful songs of war
You feel that old skin
Your old life as the marine
The soldier
The muscle memory your hands keep
Your foreign tongue and
The gun that spoke in prayer

Smoke returns to your heart
And you die once more
As you died then,

Away from love
Away from honesty
Away from home

You weep and pray for it to be quick,
Like the glorious death we all seek-
Unforgettable
Memorable
At peace
In bed
Next to your wife
Without a fear of darkness

But she does not exist when war returns
She is dead,
As dead as the dog wilting beneath the sun
As dead as the sheik covered with sand and littered
With shell casings

She cannot pull you from the confines of conflict that
Exist in your mind-
She can only weep
And pray for it to be quick

So that you may return to your heart
And breathe
Without fear of darkness-

Oh war
        Oh death

PLEASE DON'T

Please don't call me a hero, I'll break down and cry
I didn't do it for you, I believed in their lies
Please don't make me your martyr, because I did not die
Others suffered far worse, people we've come to despise

Please don't pity me for joining, because I was poor
Money for college? My soul now a whore
Please don't believe security's worth trading freedoms for
Remember to honor the warrior and not the war!

Please don't think patriotism is black and white,
Dissent is in my blood, I will stand up and fight
Please don't let us approach democracy's twilight
It's more than just paper, our Bill of Rights

Please don't listen to pundits from the Pentagon
Truth, the first casualty, the search will go on
Please don't sit back and watch it all come undone
We must all fight for change, *E Pluribus Unum*!

Please don't stop loving this country, yours and mine,
I need your help; and we're running out of time
Please don't forgive them for their war crimes
Treason and murder, these things not sublime

Please don't label me a commie, hippy or coward
My conscience my guide, peace is the way forward
Please don't leave us behind, like trash to discard
We are more than just sheep that suicide has slaughtered

Please don't just pray that things will get better
Try just a little, we can at least write a letter
Please don't trust them, not one fucking senator
They're all alike, they'd pimp their own mother

Please don't hate me if you disagree with what I say
Freedom of speech, my gift to you, don't take it away
Please don't allow this to go on; not one more day
The longer this goes on, the more we all pay.

LETTERS

After Jeanann Verlee

Dear Anxiety,
You were never mine.
Stop telling everyone differently.

Dear upper left torso pain,
If my heart is causing this abuse...
they have a hotline for that.

Dear Izod Heart Monitor,
You are not as cool as the overzealous tech said you were.
P.S. stop choking me.

Look here, Anxiety—
I was nice before.
Seriously. Cut it out.

Dear Anger Management,
To be honest, I was only with you
because anxiety was stalking my Facebook.

Dear Clouded Mind,
Now that your shit's out of my head—
I can't even remember why I invited you in the first place.

Dear Procrastination,
If I let you hang around
it just gives anxiety an excuse
to come over and fold laundry at 3:30 am

Dear Insomnia,
Sometimes we can be productive.
However please do not let me post sappy songs at
4:00 am ever again.

Dear OCD-type Behavior,
Are you high, dude?
Why did you spend two hours hanging pictures but
forget to take out the trash? Again.

Dear Overburdened Need for Security,
Why do you lock up so much if you're just going to
give away all the keys?

Dear Guilt,
Meet up for drinks later?

PRE-OPERATION

Take off all your clothes.
Put them in the bag.
Put on the gown.
Sign form here.
This might hurt a bit.
Sit up.
Sign here.
This will be cold.
Take a deep breath.
Lay down.
Relax.
Give me your arm.
Chin to chest.
Look left. Look right.
Put your arm down.
Look at me.
You will be sleeping.
You won't feel a thing.
You won't remember.

Light gray blue green
Florescent flickering hum
Peering peeking passerby
Draping curtains
Intimate exchange exposed
Sterile air shivers and nips
Charts terminals diagrams
Computer click click tap
One two three white coats
Four five six
Washed gray-green scrubs
Malingering meddling nurses
Head nods and stares
Checked boxes
Bleached blankets
Silent robotic doors
Rolling mute beds
White washed concrete
IV seeps

What is your full name?
Are you in pain?
What is your date of birth?
Are you cold?
What is your last four?
Are you allergic to anything?
What is your full name?
Does that hurt?
What is your date of birth?
Do you want a blanket?
What is your last four?
Are you feeling well?
What is your full name?
Do you have any questions?
What is your date of birth?
Are you scared?
What is your last four?
Do you understand?
What is your full name?

MEMORIAL DAY

Thank you for the yellow magnet
on your vehicle. My lifetime

membership to the Veterans of Foreign
Wars put me back three hundred

dollars. Gentlemen's Clubs waive
my cover. Shake my hand. I'm a vet-

eran. Support me. Thank me. Please applaud
when I enter the airport. When I told

my mother I had to go, I cried. Hard.
In my heyday I could bench two-fifty.

Yes to bacon on everything. I can
be buried at Arlington whenever.

Applebee's served me for free
with tip, tax. God, that girl

can sing. Chew that hot dog. Hold in
that fart. Smile. Put your hand on your heart.

WAKE UP CALL

Six days before Sergeant Justin Crossnew's
fifth deployment
our chaplain
came to say
no morning
accountability
formation
and remain
in the barracks
until our
silver-starred
squad leader
had been cut down
from the third story
fire escape
where he had double-
wrapped paracord
around his neck
and let himself hang
for the whole world to see.

LISTS & SCALES

I hate being alone with him again
Even in my mind
Trapped in his loony bin of an office
Suffocated by these four walls padded in
 frames full of bullshit
His accolades for a job well done
From people who don't know how to do their damn jobs

I can feel him staring at me
Counting my tics, fidgets, and sweat beads
Sizing me up to see if I'm a good fit for crazy

Talking to me very slowly
With small words
About things bigger than he understands
Asking me questions
And only listening for answers from his lists
That tell him how far up I am
On the "About to lose her shit" scale

We begin down the list
"Do you know why you're here today?"
Because I'm fucked up
Because I want help
And I'm here today of all days
Because I couldn't get here any sooner
Five years of wandering
Three years of darkness

And 18 months of trying to get this stupid red tape to
    feed through your system properly
18 months of waiting for someone to say they are ready
    and willing to hear my story
"Yeah," I shrug "I lost my marbles somewhere
    between here and Iraq."
He moves me up the scale

Down the list
"Do you think about hurting yourself?"
I think about how I hurt myself with each piece of my
    humanity I gave away
Avoiding eye contact with people who look like my kin
Believing in enemies
Provoking and consuming fear
The pain of remembering is far greater than any I could
    inflict upon myself
I cut myself to find relief
"I think about stopping the pain."
Up the scale

Down the list
"What about angry? Do you feel angry?"
THE FUCK KIND OF QUESTION IS THAT
    ANYWAY?!?
Do happy people come here?
Of course I'm fucking angry
I'm angry about your stupid questions
I'm angry that no one seems to notice people are dying in
    war

While everyone else agonizes, over which brand of bottled
   water to choose
And I'm pissed that I can't just go back to not giving a fuck
   about any of it
"Yes, usually."
Up the scale

Down the list
"Were you ever raped or sexually assaulted?"
Rape is sexual assault, douche bag
And your question is reminiscent of the act
Penetrating and touching me without my consent
Whatever I tell you won't matter
I know you aren't here to help me find justice or peace
Where is the question that asks,
   "Have you sexually assaulted anyone?"
How many crazy points is that worth?
"Yes."
Up the scale

Down the list
"How many times a day do you think about your triggers?"
I roll my eyes into the far corners of my mind
Into the memories that landed me in this chair
They appear sporadically throughout my day
When my bathroom suddenly becomes a port-o-john shaped
   death trap
A helicopter overhead converts my back yard into a
   flight line
Or the 4th of July turns the whole world into a place
   rated for hazardous duty pay

But my memories aren't on his list
His questions are too one size fits all for that
"A lot I guess."
Up the scale

Down the list he continues
In an effort to get to know me
And ascertain how much less like myself I have become
How strange is it to get to know someone
Whom you intend to forget

At the end
I feel at the top of his scale
The whole experience
A deployment déjà vu
Of unpacking and re-packing sea bags of crap
To make sure you've still got it
His questions forced me to open up
 my whole box of awful
Just to take inventory of its contents
But no one tells you what to do with what's inside
I guess I will just pack it back up
Haul it all home
And try not to put any of it in the wrong place

THE GOOD GUYS

"Bang-Bang," I shout as the plastic muzzle jumps, "You're dead!" I declare. "No I'm not, I'm the good guy!" my brother proclaims.

The cold steel moves across my fingers as the fence passes in from of me. I look out and watch my father patrolling the airfield.

On the television screen, I see the helicopter take off as Sergeant Elias is riddled with bullets. How could they betray him? I take a sip of coke and remember, we're the good guys.

The Twin Towers crash and burn on television while the media spins webs of propaganda. Time for us to saddle up and kill the bad man!

Stepping off the ramp of the cargo plane, the dry air ignites my spirit. Freedom fighters experience a rebirth on the tarmac and I feel one with my fellow soldiers.

My heart pounds the drums of war as I race from the helicopter, the blades beating in time with my racing heart. We kick down doors. The whining of the rotor is replaced by the shrieks of children. Don't worry, folks, we're the good guys.

Somewhere a trigger snaps and a man falls. A light bulb dims as grey matter cools. Don't stop, don't think. As I take a knee, I can't help but question if we're doing the right thing.

Room to room I push searching for what, I never was told. Under a star lit moon the boy's eyes stared coldly. You are not who you think you are. His dark pupils burned through the back of my brain. In him I saw my own innocence lost. I am no longer one of the good guys.

WHEN WILL I COME HOME

A truck drives down the country road by my house
A simple black and white dump truck kicking up
  dry earth as it passes
An innocuous event that ignites my nightmare in a
  flash
Without warning I'm there, standing at the front gate
  of Camp Griffin in Baghdad
Dry grass sways with the push of hot air and I hear
  familiar sounds of gunfire and lurid screams
I smell dry desert air tainted with camel shit and
  burning fires as fear overcomes me
I shake my head to end the terrorizing flashback
Will I ever feel safe?

Suddenly, I am back in my kitchen listening to calming
  sounds of the creek
Tears stinging my eyes as I try to fight off the
  disturbing thoughts
I become enraged by the abrupt change in my mood
The smile drops from my face as grief and anguish
  set in
The loss of a friend pounds into me as though he died
  again today
Iraqis I was unable save, echo their screams into my mind
Will my war ever end?

Looking out my kitchen window, I watch the green
  American grass melt into the desert sands of Iraq
Holding onto the counter, begging not to be sucked
  into war again
I rationalize that I am in upstate NY, now fighting a
  phantom war
My soul constantly pacing between today and Iraq
Lost between two borderless worlds blending into
  one another
Making it nearly impossible to distinguish between
  past and present
I am never fully living today, because I died there
Will I ever come home?

CHAIN SMOKING, STILL

I sit cleaning my M4
going through the motions
as I stare out at a depressing landscape
from the roof of our palace
chain-smoking and listening to Dave Matthews Band
on a hand-me-down iPod
as my eyes are drawn to a disfigured mosque
and I take another drag
and the Tigris catches my eye and memorizes me,
brings me back to the rivers of my youth
and I'm aching to swim but know that these rivers
are nearly as polluted as our minds
with chemicals and oil and death...
don't drink the water
and don't swim
but I'm drowning
every time our eyes meet
we were both taught to see monsters
but I just see pain...
and I sit here now chewing dates, drinking chai
and chain smoking, still
as I look up at the sky I don't see god anymore
and I'm sick of cleaning this goddamn M4
and god is in those 'monsters' eyes
I realize it now so I cry
and I look out at this city
with thousands of years of history
and I dread our next mission

and I kiss the barrel
and contemplate jumping off this roof
it could look accidental
but I'm a coward
so I'll play god again tomorrow for others
since I can't seem to kill myself
and for now I light another cigarette
and stare out between two rivers
puffing smoke signals to the heavens
from this biblical land,
an SOS to the universe
but who knows?
maybe tomorrow
I'll choke to death on a date pit...

RE-DEPLOYMENT

The Re-Deployment series offers viewers a window into a side of war often overlooked; life after war. The body of work focuses on my personal experience of returning home from combat. This specific photograph is about enduring trauma…not a "soldiers trauma," nor the "trauma of war," but human trauma. I believe there is a barrier between soldiers and the civilian population that leads to misunderstandings and hinders the ability for civilians and veterans to relate their trauma to one another. My work serves as a catalyst to bridge that gap and create a dialogue of understanding.

PTSDepression

I have PTSD

the docs say it makes me angry
family says it makes me distant
brain says I'm damaged
but my enemies say I'm persistent

it starts by
raising my thermostat
heat covers my body
I begin to sweat
knees buckle like broken dreams
as I fight to ignore my regret

this is my life
watching myself
a poster child for mental un-health

I hide in the corner
I can't learn to live
with this disorder

feels like
every morning's day one
this problem ain't based
on the clouds or the sun
it comes from some person that lies deep within
strangely enough
I think he's my friend

DONE

gone is the will to be poetic
to try and take those moments in time
and translate them into sounds
that make memories materialize for you and me
to try and convey what it's like
to fear for your life for a year and some change
so much so that you forget it's fear
and it becomes an anchor

I've tried to put into words what it's like
to inhale filth and death with every breath
to want to purge until nothing's left
and explain how first volleys of 21 gun salutes
make strong solders sob like children
gone is the will to make things make sense
for forgiving fools with their good intentions
convincing the country to come to attention
for repeating deeds that should never be mentioned

So... how was it... did you kill anyone...
I'm done

WHAT NOT TO WEAR TO WAR

Won't be needing that Grand Mammy Pappy Christmas
sale sweater or the zip your penis onesie with plastic
bottomed feetsies.

Drape that varsity jacket over the shoulder of the road
to war, rubber neck wreck, trophies end up in a
basement.

Holler for your new team, patches and colors,
cheerleaders under the bleachers.

Thoroughly wash your wool Mexico pullover hoody
before you stand before High Horse Six, explaining
how Mr. Pinhole Burns let down the elite unit of
fools and killers.

Before the show is over, don't be that guy wearing the
Operation Slaughterhouse VI concert tee bought at
the PX.

Save it for later and sew on your Ford Ranger Bar tab
so people know you rock.

Fall out of formation before Top notices your JV
Airborne Eagle Scout Belt Fed Canned Ammo Drive
patch.

Wear the crucifix you turned upside down and
sharpenedinto a sword underneath your uniform out
of cultural sensitivity.

Clearly display your class rank, in death, Generals wear
diapers and Privates wear portable toilets.

Sport the same smile in the ruins of the empire you
wore in the ruins of Baghdad so as to not make babies
cry.

Lace up a dirty pair of old combat boots when you
attend the war photography exhibit.

Veterans and their guests get in free.

WRONG FUCKEN HOUSE

For Nathan Lewis, who suggested it be written

I live in a ground-floor flat
facing the alley. My windows are open
to their screens and my sleep is carried
on a rhythm of bread: the drum roll
of mixing machines, break time chorus
plotting in Polish, trays on wheels
of loaves to cool, the baker's cat pouncing
her single word.
                    One night in the summer
a new sound in the alley. *At last*, you think.
You are listening to someone razor-blade the screen
next to the bed. A nostalgic knowledge
refreshes your purpose: you reach under
the mattress and draw the kabar from its
kept-unsnapped sheath. You'd bought it on post,
Ft. We-gotcha, Arizona, on one of those
garrison Fridays with nothing to do
but prowl the base and shop for knives.
It was in your crash bag on every mission.
Awled onto the sheath is a map of Iraq
and three X's.
                    In a squat under the sill
you speak out the title of this poem,
having decided that the rules of engagement
in one's own country should include a warning.
You hear a man freeze, which is to hear a man
hold his breath, then the tinny fall of his small tool
striking the concrete. He escapes like steam
and I am asleep again with the fruit
being folded into flour and the handfuls
of almonds being tossed into bowls.

AMERICAN MARTYRS

Red, white, and blue,
another folded triangle.
A rifle, some boots.
Dog tags that jangle,
like chimes in the wind,
that absolve away sin.
A lone empty chair,
at the dinner table.

Marching in grace,
boot heels that clack.
Present arms, Taps,
rifles that crack.
A well dressed soldier,
presenting me over,
to a crying mother
dressed all in black.

Children stand by,
naive of their fate.
Are paths predetermined?
Will they take the bait?
A boy steps from the line.
It's a matter of time.
Before he makes an oath,
and Death marks a date.

Medals and ribbons,
and names on a wall.
Statues and movies.
Pictures hung in a hall.
Books about history,
parades on anniversaries,
beckons the youth
to answer the call.

Driven by legend,
the glory of war.
Sharpen the swords,
beat drums once more.
It must run in the blood,
to die in the mud.
I'll never understand
what this memorial's for.

LAY IN IT

When your Egyptian cotton sheets
offer the comfort of rusted metal teeth
you find yourself contorting to sleep,
squeezing your bones into the irony
of an empty love seat.

The bird lands, kicking up a storm of sand
that settles in the deepest parts of you,
leaves an uneven deckled edge in your mind.
Nothing will ever fit perfectly again.

Watch them. Yes, you.
Watch them loaded onto the bird
again
and again.
Watch them.

A mind reloading images like your
        gas-operated,
        belt-fed,
        air-cooled,
        fully automatic
        crew served weapon.

They have been gone for some years.
These dog-eared pages are their ghosts,
the tattered stories you won't let go.

Those bunk beds are
the smashed clock of your twenties.
The gears and springs
lay across cluttered floors.
Childhood toys no longer remind me of youth—
They are priorities that lead to
a sink full of dishes, another weekend of laundry
a bedside coffee pot gives you an extra five minutes

Before waking the boys—
wake up,
skin like a wet bar of soap.
The dresser that held the linen is empty,
lay in the metal teeth.
Roll over and see the only one
staring back from a dirty vanity mirror
is you.

NO WATER AT A TRAVEL CENTER ALONG THE PA TURNPIKE

Caution signs block the restrooms
and I see why my fellow travelers are lost.
A scrap of cardboard, hand-scrawled, mocks:
no water.

The Starbucks is closed
and the Burger King staff
are outside smoking.
No water means
no one draws soda
from the colossal drink fountain,
all 15 choices dry in an instant.

At the parking lot's edge
I piss in the weeds
defiantly, as I often do
walking home following last call.
I watch children across the parking lot
hold their crotches and dance,
moms and dads pointing toward
port-a-johns strapped to a flatbed,
reverse lights squawking
like the hose trucks that spray-cleaned
the shitters around Camp Taji.
Those same third world
contractors stood atop our toilet seats
and squatted to urinate,
sometimes busted the hinges,
always left mud and sand on the ring.
"Job security," went the joke.

And I remember Iraqis lined up
behind a water buffalo's tap,
buckets, gas cans,
empty Coke bottles in hand.
Private Martin, rank-crazed and sunburnt,
used the butt of his shotgun
to turn away a touchy beggar,
busted the guy's jaw
with a deliberate swipe.
The sound was like ice
crushed inside a wet towel.
Everyone was thirsty that day.
Our Army-issued Camelbaks
held enough for just one.

Crossing the parking lot
I hear parents say it'll be all right,
you'll get to go soon enough.
But maybe it won't be all right,
I want to lean over and whisper,
like an old man waiting for death
in a nursing home.
Watch the playgrounds
crumble like Baghdad.

Alone in my car
I merge onto the highway
think about why
for fear of losing everything
I will never find love.

PARALLEL HELL

A road construction jack hammer
sounds like a 50. cal
Metal chink chink chink
as hollow casings fall

The embarrassment's the same
Flinching from out going fire
As climbing out from under a table
after the pop of a car tire

The smell of cooling barrels
leaving the dead for cats and flies
smells like the black powder
in the air on Fourth of July

Drunk before deployment
worrying if we might lose a few
is the same drunk coming home
after our worst fears came true

The darkness in my head
when I block out all the sound
Is the same calm that we feel
when we're six feet under ground

POSTCARDS

Postcards are constructed images in drop boxes. Each piece is composed of several different images layered together to create a unique scenario. In order to establish the intended perspective, each box features a prominent hole from which the viewer may peer into the orchestrated scene. Formed from photographs taken while serving in Iraq, and juxtaposed with images from the American landscape, each scene combines two familiar worlds one experiences while serving abroad. Therefore, these scenes effectively represent the imagination of an American soldier on duty. Soldiers deployed overseas remember and project their home experiences on foreign landscapes on a daily basis. When the veteran returns, he/she continues to mix the two locations in their mind. Although the returned veteran appears to the rest of society like any other person, the soldier remains intertwined within two realities.

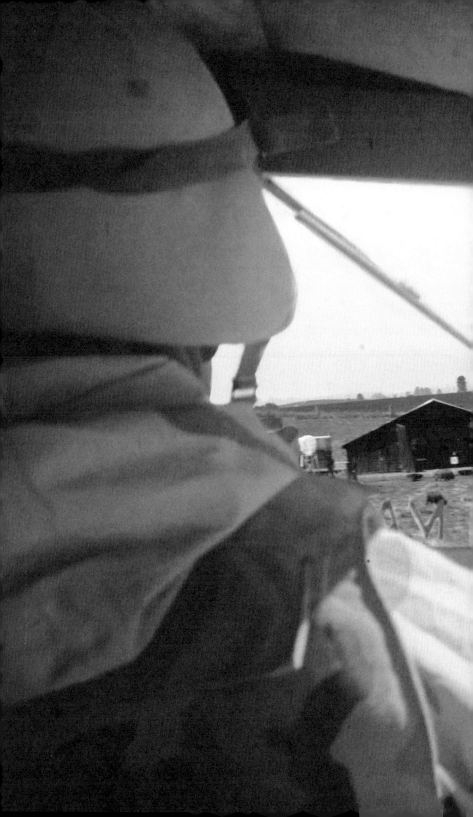

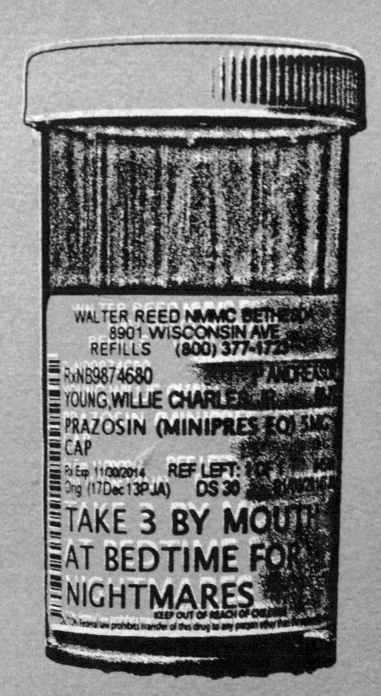

DIESEL TRUCK TIME MACHINE

Sunny skies
Walking to class
A skip in my step, a song in my head
Wave to friends
Search for glimpses of the river showing through the
 houses

Pierrepont Avenue
An old route slicing through the setup
Amidst the folly and flop
Tremendous party houses hold out another year

My footsteps
Two fall per sidewalk square
Step on a crack get sent back
A truck drives by and takes me to Iraq

The sound, the smell, especially the sound
A guttural rumble
The distinct sound of a large diesel with its engine
 brake on
After the noise the exhaust blows through
Both senses store memories

Carried away briefly in the fumes and sound waves
Every bit of conciseness
Gone, Zap
All the way back through time
Across oceans

Back to Iraq

Guthrie's driving and I'm riding shotgun
Its late morning and we're lost
The officers got us lost again
Taking an exit the road disappears
Simply disappears

Tires and trash are burning
We announce our presence with a cloud of dust
Barely enough room to turn the convoy around
People are running from us
Tank columns blasted their way through here weeks ago

A tall man holds a shovel and is standing above two
  graves
The child clinging to his leg buries his face behind trousers
Don't stop here
I point my weapon and finger the trigger

He points back with an accusing finger.
Points at the graves
He makes the throat cutting motion and points directly
  at me

Rubble is blasted all over
Huge chunks of concrete and steel
Something in the ditch is dead
You can smell it
Spent shell casings in the dirt lay out like
  the yellow brick road

Ghastly evidence of a crime
It's everywhere and you can smell it

A few steps later I'm back on Pierrepont Ave.
Looking for friends
A golden retriever drives by with his head out the
 window
and smiles at me
My heart beat returns to normal

I don't have to go down this street
Sometimes I go different ways
There's always other ways
But sometimes I take Pierrepont and intentionally
 drop my guard
Wait for the inevitable reptilian brain animal panic rush

Take the route and go back
Back for one minute
Just for one second
Just for a thought
Just for a memory
The rubble, the smoke, the man with the shovel
I visit them in memory so they don't visit me in sleep

FIRST-CLASS GUILT

On a recent flight home from a veteran friend's wedding in Honolulu, I sat in an aisle seat in coach. Two college-aged boys sat beside me. They had long, muscular legs and necks wider than their heads. Maybe they were football players, wrestlers. Why they were large doesn't matter—what matters is I knew the flight would be more uncomfortable and cramped sitting next to them. As we taxied, a male attendant walked the aisle asking repeatedly, "Do we have any military families on board?" No one replied.

Although I thought I knew what he meant—veterans, or spouses or children of veterans—I didn't say "yes" or raise my hand. The question was vague, and I didn't know why he was asking. (Also, being a former soldier, I am always hesitant to volunteer. You see, in June of 2001, when I was 17, I volunteered for six years of service in the Army National Guard. I ended up spending 11 months in modern-day Mesopotamia. I've been more careful since.) The attendant walked to the back of the plane and then returned and stood near the security exit a few rows up. I overheard a conversation he was having with a passenger about how some of the first-class seats were open and they were offering them to veterans or their families. After a few minutes, he walked down the entire aisle of the plane asking the exact question again. I didn't say a thing.

It's not that I didn't want the seat. Maybe it's my personality coupled with the burden of the identity each person takes on when they become a "veteran," but I didn't want to announce that I was a veteran to a bunch of strangers on a plane. In situations with friends or family, it can be awkward when someone announces, "Yes, Hugh here is a veteran" or, "He was in Iraq." In almost all instances, either no one knows what to say or they thank me. Obviously it differs from veteran to veteran, but my preference has always been that if someone wants to sincerely discuss Iraq, soldiering, the war, I'd just like to talk one on one, not spontaneously in a group setting where I usually have a few minutes to sum up my experience and explain my version of the war. In the presence of strangers however, the brief looks of attention are too overwhelming and although I truly like discussing the war, soldiering, and so on, I prefer not to have to do it with a stranger on a plane as an audience overhears everything. (I will admit that this "issue" of when and where I discuss my service is incredibly minor. Just ask a Vietnam veteran how they were treated after returning to the States.)

Sitting in that seat, I imagined what to do. I could be the loud, motivated type and raise my hand: "Sir, I am a veteran of Operation Iraqi Freedom Two reporting for a first-class seat." Or the more direct "Yes, I am a veteran." Or maybe the proud "I served honorably in combat in the war in Iraq." A logical person might ask, "Why not just relax and ask for the seat?" It seems simple in retrospect, but there's not much related to the Iraq War that's been simple for me.

* * *

The plane took off. After 20 minutes or so, as the same

attendant began distributing drinks, I leaned out and held up my hand to stop him. He bent down, attentive.

"I have a question."
"Yeah?"
"Did you say there's seats up there? I'm an Iraq veteran—"
He cut me off. "Oh yes. You should've said so earlier. Come on up."

He waved for me to follow and led me toward the front of the plane. When the curtains separating coach from first class parted over my shoulders, I was relieved to make the transition so subtly. The attendant pointed to a back row of first-class seats, and I sat in the one by the window.

"Fish or roast beef?"
"No, I don't want to buy anything."
"It just comes with these seats. You should get something."
"Okay."
"The fish or the roast beef?"
"Roast beef."

After a few minutes, the attendant set a tray with a blue placemat in front of me. A fluffy roll rested on a white doily beside the plate. Steam rose from the roast beef wedged against a mound of mashed potatoes drenched with mud-colored gravy. I thanked the attendant.

Because I was briefly involved for just under a year in the Diyala Province in northeastern Iraq, this warm plate of roast beef awaited me. That was why the seat was so wide and why I had all this space to myself. But I had deployed only once to Iraq. When so many others, including friends of mine, had suffered two, three, four, five, or more deployments to Iraq and Afghanistan, why should *I* be the one enjoying this seat's comfort? Given that less than one percent of Americans have served in these wars, it seems laughable to feel guilt over this fact, but it's there, always fluctuating with my mood. I can complain about the tight space back in coach because I still have a set of arms and legs; many others who've come back from these wars can't say the same. Others haven't come back at all.

* * *

I justified taking the seat not as an "award" or "perk" of being a veteran, but because of the practicality of it: I'd be giving more room to those larger men in my row. It was the polite thing to do. It'd make everyone's ride more comfortable. This was not a way to pat myself on the back, but to create a concrete, legitimate reason why I should take the seat for reasons other than being a veteran, being in a war. Raising my hand to inform someone I fought in a war in exchange for a wider chair with more legroom feels, to say the least, strange.

During this war, we've had 4,804 coalition deaths and over 100,000 civilian deaths—there are so many different estimates and speculations on the number of Iraqis killed during this war that it's hard to find an accurate count. We have tried to comprehend the number of dead in many ways. The Eyes Wide Open exhibit, which began in 2004, displays a pair of empty boots for each soldier who has been killed in Iraq; the

Iraq Body Count Exhibit traveling memorial has placed over 100,000 white and red flags, representing Iraqi deaths and American deaths respectively.

One memorial that tries to communicate the effects of war, at least on the American side, is the Vietnam Veterans Memorial Wall, which lists the names of all U.S. service members who died or went missing in action in that war. The experience of seeing the names along with your own reflection in the black gabbro stone is one attempt to try to understand that many deaths.

In Larry Heinemann's novel *Paco's Story*, a minor character named Jesse has a different idea of what a memorial should do. A Vietnam veteran, Jesse imagines a "Vietnam War Monument" that would be located on "a couple acres of prime Washington, D.C., property..." and made "with Carrara marble...the whitest stone God makes." He then explains the remaining details in a long-winded tirade over coffee and a bowl of chili:

In the middle of all that marble put a big granite bowl... about the size of a three-yard dump truck...Collect thousands of hundred-dollar bills, funded by an amply endowed trust-fund, say, to keep money a-coming. Then gather every sort of 'egregious' excrement that can be transported across state lines far and wide—chickenshit, bullshit, bloody fecal goop, radioactive dioxin sludge, kepone paste, tubercular spit, abortions murdered at every stage of development...shovel all that shit into that granite bowl and mix in the money by the tens of thousands of dollars...Then back way up and hose down the sod.

He suggests Americans could come and "take off their shoes, roll up their trousers" and wade through all of it, the reward being that they could keep any of the money they found. Although Jesse's suggestion could obviously—I imagine, at least—never happen, the argument behind this "memorial" is clear: we sugarcoat war and rarely touch the blood, shit, and death it produces. We Febreze war like it's the inside of a gym locker.

* * *

It would be melodramatic to say I stared at that impeccably presented first-class meal and thought of any of this. I didn't at all. Somewhere though, in my subconscious, I felt a strangeness, a Beckettian absurdity. The row to myself, the absence of others, the curtain that separated me from coach—all of it reminded me too much of the gulf (no pun intended) between civilians and the realities of these past 10 years of war.

On a superficial level, I did enjoy the physical space to stretch out and relax, but it was also the comfort and quiet of first class that made me sense the guilt that might not have surfaced had I stayed in the claustrophobic coach section. Sometimes mild and sometimes extreme, it's a guilt I confront each time someone thanks me or offers me a perk for my veteran status like a free seat, a complimentary meal, a parade, a round of applause at the airport. But I tell myself: We as veterans need these small, sometimes insincere and shallow acknowledgments, because otherwise when would we know the civilian population even considers our war(s)? Paradoxically, although I know this treatment makes many veterans feel respected and welcomed, it sometimes seems to me laughable and pathetic to participate in a war where hundreds of thousands (if not more) died and then be greeted with a wider seat and a warm meal.

Many times, just inside my head, I've argued that we need fewer of these gestures aimed at veterans—free meals, free upgrades, parades, applause, even *thank yous* (which are often sincere)—and more, at the very least, reflection on and acknowledgment of the horror and death soldiers and civilians in Iraq and Afghanistan experience. I think most veterans would prefer to be "honored" and "respected" by having fewer issues with the VA. By having a civilian population that's more informed about the nuances of these wars. By encountering fewer people who are, or just seem, indifferent. By being sent into other countries with more careful consideration and planning (books like *Fiasco* by Thomas Ricks and *State of Denial* by Bob Woodward, illustrate this better than I can). The treatment of veterans has improved since Vietnam, but there is ample room to do better.

In W.H. Auden's poem "Musee des Beaux Arts," the speaker discusses the story of Icarus, who he sees in a painting by Breughel. As Icarus flies too close to the sun, his wings melt, and he falls from the sky to his death while "the sun shone / as it had to..." and "the expensive delicate ship...had somewhere to get to and sailed calmly on." The suffering happens, as all suffering happens, "while someone else is eating or opening a window or walking dully along." Just as our wars go on, those who aren't close to the threats or close to those going overseas will always have trouble feeling a sincere concern for the suffering that takes place. But many veterans and Iraqis can't end things, like most American civilians, by turning off a screen.

The longer I am away from my experience in Iraq, the more complicated it becomes. When people ask questions—"Should we have been there?" "Did you see *progress*?"—a "yes" or "no" always oversimplifies the situation. Maybe I avoid discussing it because many times I feel that whatever I can say, even the concrete stories I have to tell, just can't ever come close to doing the war justice, to telling the truth about something so complex, as entangled with multiple perspectives as the enormous webs of telephone wires we drove by while on patrol.

A yellow ribbon, a parade of bright flags on Memorial Day, a wide seat, and a free meal—all of it seems too neat, too hollow and futile, in comparison with the suffering and death war brings, specifically these last ten years of war. Yet I didn't move back into coach. I didn't send back the food. Eight years after my time in Iraq, I hate that I myself am one of those people "eating or opening a window," or in this case, flying "dully along…"

In Dostoyevsky's *The Brothers Karamazov*, Kolya says, "It's funny, isn't it, Karamazov, all this grief and pancakes afterwards."
Or roast beef.
What could I do? I was hungry. I cleaned the plate.

I'LL CARRY THAT FOR YOU

Dear friend, don't worry
I'll carry that burden for you.
Place it on my shoulders
All I ask is this; Never forget me.

My friend I'm happy to carry your freedom.
Thank you for believing in me.
What's that? I'm a hero?
I never thought so.
Just doing my job.
I do it so you don't have to.

Friend, where have you been?
Busy? Oh, Sorry to bother you.
Where did your yellow ribbon go?
Okay I understand.
Yes I'm still carrying it for you.
It gets heavier every day.

Friend! Friend!
Where are you?
Oh there you are.
You're hard to find these days.
What's that? You have to go?
Wait!

Friend,
I carried that burden as long as I could
Now I'm broken and hurting.
I could use some help.
You are nowhere to be found.

I DECLINE

The Buddha
once told of a foolish man who,
knowing that Buddha always
returns evil with Love,
went to Buddha and
administered a great abuse.
Buddha pitied the man
for his mistake asking, "If a gift be
declined, to whom does it belong?"
The man answered, "The gift belongs
to he who offers it."
Buddha declined the man's abuse
Requesting instead that he
keep it for himself.

As a child, I knew not of the Buddha's
lesson, so for many years have
carried your abuse with me.
I tell you now, I decline to accept
and leave it instead for you:

"I think you're a fuck up."

DROP YOUR PACK

it is time to drop your pack, youthful spirit
your back is busted and you have too many years ahead
to carry these burdens upon your guilt stricken body
time to walk about with concentration on now
because that was then and this is how you move forward

one step at a time towards forgiveness and compassion
doors are always opening but you know
that key cannot always be trusted to let you in
this character has been deflated enough to know where
arm's length ends and vulnerability begins

maybe it is time to put down that weapon for good
it has taken many forms throughout the years
its resting place has become dusty in its absence
the blade is weathered from battling the past
its clanging cold steel creating sparks in the night

hands and face are scarred beyond years and you are tired
hardly feeling, just surviving, and losing stamina
approaching each interaction with an elevated level of
  threat
blade still deadly and becoming increasingly cumbersome
a sharpshooter's eye does not focus like it once did
but the precision instinct remains
continuing to follow tracks through the woods and
  gaining speed

all the words from past stories
are vague and blurry and difficult to read
let us sit silently, because this is now
and honesty only gets lost within words
I share your vulnerability when I look into your eyes
I've just been pulling out the sadness and replacing it
  with madness

waving goodbye to those archaic tales
transgressions of a former you
will continue to be present in shaky bones
if you cannot find your way home
then there is something missing inside you
let it go, forgive yourself, forgive others
release any hope for a better past

I HAVE BEEN

I have been the child, rapt in books
awakened by strangers, words like
laughter obscuring banter, waiting
for the music to fill me up

the girl, so trusting, so naive,
saying it was wrong for us to go
because it wouldn't change a thing
war has always been a thief

the almost soldier, breaking ranks
crossing lines, burning sonnets
in the shadows of a conex full of
slave cables and moldy half-tents

I have been the young woman, curled
over, chest to knee, recalling friends
gone, smooth ache, blanched face
and homesick for the war

the sergeant, so careful, so at ease,
saying it was hard to come home
because I couldn't change a thing
war has always been a thief

the veteran, zip-tied and bleeding
in a cool, cement cell, letting go
bare feet, long breath, waiting
for another door to open.

ON THE EVE OF SEPTEMBER 11TH

On the eve of September 11ᵗʰ I held my friends' new baby
awake and against my chest, our hearts collide.
little left hand cups around the interior point of my
  clavicle,
where the two bones come so near together,
at the bottom of my neck, and I am learning to breathe,
again.

And my body hurts.
it has been aching for three weeks now with those war drums
drumming, sound bites that sound
like the same old shit all over again.
And it hurts. It hurts my body, and it hurts my friends.

earlier that night a bug flew inside and it was –
making a racket – in the corner of the living room
so, like any good marine, my friend, the father –
he grabbed his drum sticks and we
proceeded towards the corner and that side of the room
not knowing if the bug was in the corner, near the ceiling
or hiding somewhere in the blinds or by the windows

we tap all around the lights and windows and walls –
and look up and down,
after awhile we turned this little charade
into a performance art piece, call our troupe the:
"*stoned jarheads*"

we finally find the bug in the light
my friend says he will lower it down
I can smash it with my shoe

and then my friend decides, instead of smashing it
we can just take the lamp outside
instead of tricking the bug out just to kill it
and then it could just fly away out there and simply, be

this all happens in about five minutes
how simple it was not to make any quick decisions
limited strikes, at the bug, and let it go, peacefully
it seems this has taken other men thousands of years,
and they still just want to kill whatever it is
that bothers them

to just kill what they don't understand.

On the eve of September 11th
I hung out with my friends, and their newborn baby
we ate ice cream and drank coffee with maple syrup
and coconut creamer –
we had delicious lasagna
and babies spoke softly without words
with infinite knowledge

do you smell, *that smell*,
the mother says – and then says to smell it's little
baby head
This is where some magic is happening.

On the eve of September 11th
We didn't talk too much about it being
September 11th

we talked a little about war drums
and our bodies, and babies

the same shit keeps happening over again
and it hurts my body
literally, and in this body
I didn't even fight
in a war

So on September 11th's I just talk real talk
or sit alone in silence
And on the eve of justifications of war I must do things
which are life,
affirming

I went to therapy on the 11th this year
and afterwards the chain on my motorcycle busted
and broke the transmission cover open
a new friend now sweetheart came to get me
took me to a pond so I could float in the water
like my maternal instincts told me to
it will be about $600 to fix this mess
and $900 for everything to get into working order

I ain't really stressed because I've been so full of
suicidal ideations –
I just know that for a while, and even if life is a while,
that's just part of the terrain

so if you see me stranded on the side of the road
with broken chains, or a broken heart
just take me to the river, or a pond, and drop me in.
Lay me down softly. Slowly.
Guide me gently, hold me sweetly.

I have priorities. War is not really one of them

I believe in getting from point A to B, if and only if
I can
I believe little humans are, like my other friend says
our tiny ancestors, with knowledge to share
I have no time for propaganda,
and I swear jobs are triggering
sometimes just because, it's a job
I have some insight
and war is not something I want to support, anymore
I believe that people do bad things, and that
it is often done by using other people, to do those things

so when I can – when I can I hold babies
and when I can I drink coffee.
And get a little stoned with friends
I remember these war drums, these
war cries –
and even when I try to ignore them they still affect me
today

My body hurts. And I keep crying.
I do not really know if it will all be
okay, but I have this coping mechanism
where I tell myself it will be,
okay.

And I can laugh, and cry, and get mad
and do it all over again

on the eve of justifications of war, which is sometimes
everyday
in America
I do something that is life,

affirming. Like holding a child
Holding space
Letting go
I grab coping mechanisms and started to
write my heart out,
and my friend, asked me what it would be like to just
write into it

To just write into it. To get right, into it

On the eve of September 11th I held my friends'
new baby
awake and against my chest our hearts collide
I wear a shit-eating grin
and I am so high from this life force pressing against
mine
Little left hand cups around the interior point of my
clavicle
where the two bones come so near together
at the bottom of my neck, and I am learning to breathe,
again
I am learning to breathe, again

And on the 11th, if you see me stranded
on the side of some road with a broken heart,
or broken chains
be gentle, share with me your, favorite poetry
Tell me stories of babies and stoned jarheads
and cacophonies of peace and help me
replace those war drums with heart beats and I

I'm gonna get right into, just write into it
On the eve, of justifications of war

give me something life affirming
and on those anniversaries, on those
worst-days, if you see me stranded
on the side of the road
with a broken chain, or a broken heart –

Just take me to the river, or a pond, and drop me in
Lay me down softly. Slowly.
Guide me gently, hold me sweetly.
I am learning to breathe, again.

DRAGOSTE

The woman was my Rati, my Eros,
my Aphrodite, and my Frejya. Until
she was not.

The lady was my romantic juxtaposition,
a struggle between
love and lust, persistent but guarded, sacrificing
ego for embrace. An unconditional union.

But, me, I am
war wired dysfunctional, flaunting
intimacy inadequacies, leaking laments and
leaving zero space in the expanse that we
were.

I'd like to think I warned her.
I'd like to think she knew. But,
the war breathes sometimes for me,
and she must have thought, at some point, that
we could only be us,
for time less acceptable than she could have known.

rEVOLutionary Meetings

I once read a love poem at a revolutionary meeting

You bring the day
tiny toes down the hallway
to tell me, "good morning, Mommy"
When you sleep I count your breaths
to make sure that I'm not dreaming
But if I am, I'd prefer never to wake
for fear you would stop coming true
Love like this feels impossible to hold
So I'll snuggle you while I can

"Wait, Sista," they said, "Don't you know this is a revolutionary meeting? What about the man?"

Oh, yeah
The man

I want to love you for
your perfect imperfections
Each crack filled with room to grow
Reminding me that you are trying to be a better man
So I will always be here beside you
Supportive when you try
Patient when you struggle
Loving when you cry
And gentle when you are temperamental

They said, "Sista, that ain't revolutionary"
"What about the struggle?"

Oh, right
The struggle

We drifted away from love
Sometimes stopping to dance a familiar dance
Of goodbyes without hellos
Intimacy reduced to shared coffee creamer
We made love this morning without kisses
Shared a bed the night before while worlds apart
Growing more weak with each week of drifting past
Trying to float above the tears

"Um, Sista, you can't be weak in the revolution
Haven't you been to a meeting like this before
Sista if you ain't down, there's the door"

I couldn't take it anymore
I told the revolutionaries
That the only man we need to talk about
Is the man in the mirror
And if we want to see revolution reflected there
We must struggle to remake ourselves in love
as bottomless as a mother for her child
Full of the empathy we give life lovers
And strong enough to endure the inevitable storms
I reminded the revolutionaries
It can be hard to see sometimes
But if you look closely you'll find that
Love is inside of every revolution
Showing us that it can be reshaped

They once asked me to read a love poem at a
    revolutionary meeting.

NOT THAT KINDA GAY

"I am cool with gay dudes
So long as you don't hit on me"
Don't worry bro; I am not that kinda gay

You can't be afraid of a dick
It seems your stick is the star of countless
  comments and gestures
Your cock-fearlessness is impressive
Perhaps it is to avoid you feeling the feeling
You imagine so many girls do
As you whistle and shout
Thinking that without a doubt
She must be flattered
Why then, would you avoid that
Feeling of sheer joy and lust
That surely consumes your innocent prey
From day to day

I am not that kinda gay
I will be polite and show my manners
Rather than letting my affection come down
  like a hammer
I will spare you the
Embarrassment and shame
That comes from what you call a game
Consider these women as I consider you
Who have a mother, brother, daughter

You are confident and secure
A battle hardened veteran not afraid of anything
Except me
Bullets don't terrify you but I do
Let my lisp and gait succeed where rockets have failed
We walked through unspeakable horror
but my sexuality is more of a nightmare
But oh wait; I am no threat to you
I am not that kinda gay
I offer the humanity you deserve
But maybe I shouldn't

DRIVING TO YOU

As I drive down the road, IED's and snipers fill my mind
I check my six, is that car following me
I'm coming up on a choke point, my eyes scan
Out of the corner of my vision, I see something red
I look right and see the stuffed rose with a small frog
    sitting in it
My mind immediately shifts, my grip on the wheel loosens

I'm not at war, not anymore
My only fight is against the road, and against time
I push and push to get closer to you
With every mile and every minute I come that much closer
Like a madman I fight, pushing myself harder and harder
My mission is to get to you

On my way, I interview another veteran, the rush of war
  returns
The camaraderie, the pain, the struggle
Side by side, fighting wars far and near
But then I remember my new objective
I no longer linger, I push through the haze of yesterday
I return to my struggle with the time and space away
    from you.

There are no more bombs
There are no more guns
There are no more bullets, and for that I am thankful
While I still hear them, haunting my memories
Your love brings the light of a new day
There are only roads, that eventually lead me to you

STRENGTH IN VULNERABILITY

I have blue hair
I wear dresses
People ask me if I am a veterans' girlfriend or wife
    because I advocate
And care so much about veterans' issues and rights

I am a female combat veteran with PTSD.
But now, in an interesting twist of events
I am a step mom
Of a nine-year-old girl named Juliana
And a seven-year-old boy named Jaxon

I have a service dog named Boo
And a rescue dog named Moxie, also with a touch of
    the PTSD

I have days when my pain overflows onto the kitchen
    floor while
I'm making lunch for the kids
The dogs gather at my feet to comfort me, to brace
 my fall

We tell the children I cry because
"My heart hurts from the war."

Sometimes I sit alone and rock softly in the yellow
 chair
My stomach aches
My back screams
Boo lays his head in my lap

Moxie wiggles her tail, unsure

Juliana walks up the stairs
She recognizes the look
She walks over to me and brushes my bangs away from
  my face
And says,
"Everything is going to be okay."
And I believe her

Jaxon says I am 65% man
Because of the way I drive like his Dad
Because of the way I talk (curse) like his Dad
However, I am hot

One night I woke from a nightmare
Especially horrible since I had not had one in so long
As if time had added to the potency of the flashback
Jaxon was sleeping with us
I got up with a jolt and went into the living room
To breathe, to write, to calm
I could hear Jaxon whisper to his daddy,
"Doesn't Jenny know dreams aren't real."

I cried harder
Unfortunately, my nightmares are real

Juliana says,
"Jenny acts like a big kid."

I laugh a lot since I met the kids
And swim in the pool watching over them

And some moments,
I'm just me
Not my PTSD
I'm just me that loves my dogs unconditionally

I'm just Jenny
The stepmom with blue hair and tattoos that laughs
  a lot

ANDY KAUFMANN *PTSD* | 253

THE TRAUMA OF LOST LOVE

Fear of love is such a cliché male fear...whether it is framed as fear of commitment, fear of settling down or fear of change. Like any other cliché, it is complicated, meaning different things to different men. It evolves and manifests itself differently over time. Each manifestation can be defeated, but the seed of fear will always remain. I cannot speak to others' fear of love, but I know where mine started.

When I was 23 years old, I had not yet been in a relationship with someone I loved, fathered any children, or had my heart broken. My only fear regarding love was of not finding it. I was in college but had little idea what I was going to do with my life. Between my training and activation stints in the Army National Guard and my own failure to be a responsible student, I was a 23-year-old freshman. Every day, I thought I was a complete failure that would never graduate college. Then my unit was sent to Iraq.

I spent my tour in Baghdad as a humvee gunner. My squad escorted a military captain and four State Department police officers to police stations in Baghdad to train and support the Iraqi security forces. Baghdad, for all its troubles, made me feel alive again. Each day had a purpose, and my actions, for better or worse, had consequences. The year was not a cakewalk, but it opened my eyes to the world around me.

Some of the best days of my life occurred inside the compound of the Iraqi Traffic Police, where my squad escorted our "packages" at least twice a week, sometimes more. Early on in our deployment, we saw a dirty-looking kid walking through the compound with a large sack slung across his shoulder and convinced him to visit with us. His name was Ali and he was about 12 or 13 years old. The bag contained aluminum cans, and he was only in the area to find and collect cans, he told us with body language and a few broken-English words. We joked with him, gave him our cans of Rip It energy drink, and a couple bottles of fresh water. I hoped to see him again.

The next time I saw Ali, he was with his best friend, Ahmed. Ahmed seemed to be a little younger than Ali and more personable. Ali had to be convinced to trust us, but Ahmed was different. Over the next several months, the boys became my escape from what was happening around us. It was 2006 and the al-Askari Mosque had just been bombed, sparking the civil war that ultimately prompted President Bush to institute "the surge." I emailed friends and family at home and told them about the boys. I asked for care packages for the boys on my MySpace blog, and the response was incredible. Over the months, we gave the boys food, clothing, toys, school supplies, shoes, footballs, (American) footballs and candy. To a cynic, it is no wonder the boys loved us. It was different though.

On most days we played Rock, Paper, Scissors, kicked around whatever ball or can was nearby, or spent hours making jokes in the way that only two people speaking different languages can. We posed for photographs and talked about our families. Ali and Ahmed asked me just about every day when I was going to develop the photographs so they could have them.

* * *

One day we arrived and Ahmed was not there. Ali was sullen and I knew something was wrong. Through our interpreter, he told me that Ahmed and his mother were in line at a gas station when a suicide bomber detonated. His mother died immediately, but Ahmed, who was holding the gas canister, was alive and severely burned. I remembered reading in *Chasing Ghosts* by Paul Rieckhoff (founder of Iraq and Afghanistan Veterans of America) that, after the invasion, Iraqi hospitals would not administer care until a patient had paid, so I went around to the members of my squad asking for money for Ahmed's care. Soldiers in a war zone do not typically carry a lot of cash, but we gave what we could. Then we waited.

Our mission did not take us back to that police station for four agonizing days, and, as I saw Ali approach my humvee, I knew: Ahmed died. Ali dug a hole in the dirt, pointed to it and said "Ahmed, Ahmed." Then he covered the hole in dirt. He repeated this several times. I was crushed. Ali and I sat on the curb and cried: him, the Iraqi boy who had lost his best friend; and me, the U.S. soldier in full desert camouflage, body armor, combat boots, helmet and rifle. After a time, he said he had to go and I was left to cry silently in my humvee without him or Ahmed.

Within a week, my mother sent me a photo book of me, Ali and Ahmed together. Ali's eyes lit up when I told him that I had it. He hurriedly opened it up, turned to the first page and saw a picture of Ahmed smiling back at him. That is when he fell to his knees sobbing. It is still the second-worst day of my life; the worst was the day I learned of Ahmed's death.

I still had many months left in country and Ali and I spent them in much the same way we had before. Eventually, however, my time was up and it was time to go home. My last day with Ali was difficult. He asked if he could come with me. He asked why I could not just stay. He shouted. Finally, he burst into tears and ran away. I was heartbroken for the first time in my life. I went home soon thereafter.

Home was not the same. Like many soldiers, I avoided talk of PTSD, but something was wrong. I thought about Ali and Ahmed all the time. I realized that I loved them like sons or little brothers. They were the only two people I had loved like that and they were gone. I wondered if I would have been better off not befriending them at all.

Thankfully, fear of love and hatred of love are not one and the same. Eventually I met a woman who saw me for who I was and loved me anyway. She listened to me on the few occasions that I made myself discuss Iraq issues with someone, and she kept quiet when she knew that "I don't want to talk about it" really meant that I did not want to (or could not) talk about it. She was wonderful, and I knew that I was in love.

It was not long after that when the nightmares started. I was no stranger to nightmares even before Deanne. I had dreamt about Iraq often. I second-guessed myself. I relived moments. I woke up in terror. Those weren't the dreams I had sleeping next to Deanne, though. Now my nightmares were about life stateside. They always included something terrible happening to the woman that I loved: She died in a car accident on the way to see me. She was murdered and I was not there to save her. She was gone, and I was heartbroken again.

Deanne began censoring our choice of movies. Scenes where children died caused me to break down. Seeing wives or girlfriends die on screen sometimes did the same, although then it was more internalized and caused traumatic nightmares. Finally, she asked me to seek help from a professional.

After years of counseling at the local VA center, I can say that I fully recognize my fear of love and its consequences. My nightmares are much less common, and I am better at avoiding movies that make me miss Ali or trigger my grief for Ahmed.

And Deanne and I got married. She still knows me better than anyone and makes my life better every day. I never feared having my heart broken via breakup but my experiences in Baghdad cloak my love for her with a fear of losing her suddenly, like I did Ahmed. I still have bad dreams about losing her to violent crime or a car accident, but they are becoming more and more infrequent. And, when I do have them, they are usually clustered around triggering events in my life. In that regard, they're more manageable now.

We live in a safe area and Deanne no longer works nights. Life is good.

Before I know it, I will be starting my career and making enough money to support a family. I am still terrified of losing a child. But, if I had not faced that fear before, I would not have my wife. I expect to have flashbacks and nightmares when I have children of my own, but, like anything else, they can be managed. Love is too grand not to at least try. At some point, you have to brave it and hope for the best.

PERSONAL MYTHOLOGY SERIES

I have spent the past five years researching military veterans' experiences and making art around bridging the disconnection between veterans and contemporary civilian society. Through giving physical form to my own experiences of war, and through social engagement art projects, my work aims to create space for and facilitate intergroup dialogue around current, divisive, socio-political issues. These works are always autobiographical and frequently investigate ironic inconsistencies in my experiences of war, of military culture, and of gender and sexuality constructs. My art is caught-up in internal struggles and is often uncomfortable to build or view, but works toward honesty.

A common thread throughout my work is the idea of embedment—the relationships between veterans and their new, civilian communities, physical and invisible wounds embedded in veterans' bodies, and actively embedding a consciousness of veterans' issues within the larger public.

Daddy's Hand, p. 261 (top)
Uncovering My Crime Scene, p. 261 (bottom)
An Unfit Effect, pp. 262-263

A PRE-SUICIDE CONVERSATION WITH SPIRIT

I sing Goddess in the face of my ego,
learning over and over again, I am no hero
She says, "you are an explosion of ecstatic magic on
the brink of extinction"

I said, "whoa, yo, that is pressure beyond reason"

yes, you're a big deal
no, we won't survive without you

please, for the sake of our grandchildren
lift your dharma and breathe it in,
deeply

remember how it felt to be born
imagine how it would feel to die before your time

how many songs unsung or
oxytocin fueled jewels unwon because life got rough?

sometimes our best is not enough
and in those moments we gotta put our faith in
  something beyond us
like Love...
that juicy tender stuff that breaks all boundaries and
  raises vibrations up...
Love.
the fire that burns through illusion when fear chokes out
  the Sun...
love
infinitely indiscriminate in its ability to connect us all as One

love
the stuff lucid dreams are made of...

vomit if you need to, spit out the sadness
I know we all experience some babylonian madness...

but our presence is requested in this war
  on consciousness
without our warrior force there would be no one
  left to battle the abyss

and we need to fight, not with bullets or rage,
but with limitless Light
right is relative, and we know this,
so just stay alive with me tonight

if it doesn't serve God then its a fraud
if you can't see the shine then stop hiding in the fog

if your world is turned to rust
remember we are forged from the strongest Stardust

so when your glow starts to fade
and your embers are lost in the rain

remember...

you'll never be alone as you wander through the dark,
because we are gasoline, baby, *you* are the spark

now go blow some shit up...
but do it with Love...
and know that your mere existence in a state of
  elevated awareness is more than enough

DIGNITY

I still remember my first meal: It is seared into my synapses.
I was wearing my cashmere Saks 5th Avenue coat.

Only I knew it had come from the clothing donation room
at the New England Shelter for Homeless Veterans.

The staff waited on me as I imagine they do at the Ritz
(Which I somehow know is now called the Taj).

Calling me by name and offering extra-ordinary hospitality,
it was like nothing I could remember.

None knew what I had paid. I think I had paid *something*,
(out of embarrassment that I could hardly afford anything).

I kept sobbing intermittently, chin to chest,
between taking photos and posting them to Facebook.

Not eating

Michael had greeted me at the entrance like a friend,
then stopped by to offer bread and pastries, saying:
  "Catch me on your way out!"

How could he see through my armor of Italian cashmere,
with its patina of class inequality?

Maybe I deserve love.

Maybe I deserve happiness.

Maybe I'm worth more than 1st quarter results,
or my last quarter for laundry.

It took a corporate-funded nonprofit
to convince me that I had innate value as a human being.

Panera Cares®
So all may eat with dignity

(un)clothed and in her right mind

At times I've felt disjointed and disconnected from myself. Much of my identity was based on rank and status in the military. Rank and my uniformity were the only things protecting me from being weak and objectionable and created a body only acknowledged by symbols and insignia. After living out of uniform for a few years, I began to feel unknowable even to myself. In stripping away the symbols with which I identified, I found myself bare and vulnerable. But this same vulnerability allowed me to begin to see who I was outside the context of symbols. Through this series of self-portraits, I am trying to understand what frames identity and learning to recreate my own.

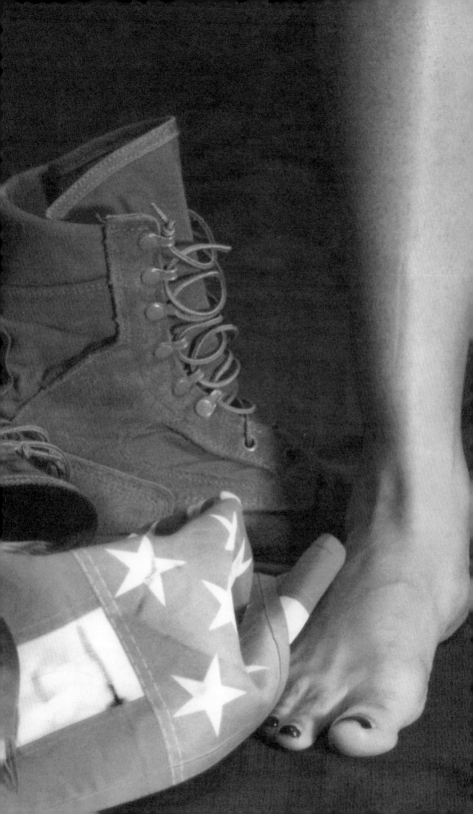

THE UTILITY OF PAIN

I inhale the hatred,
Pause to let the power seep and solidify into my heart
that sharp pain that is channeled and repurposed
I exhale a stronger, less brittle man
Nothing can occupy me so easily
Anger is my octane
The demands of war call for a life of violence
opposition of human nature
A single blemish, anything other than "Army Strong"
suggests a fatal defect
A good Soldier will do what must be done
Hard on the engine, but unchallenged in its ability to
overcome any obstacle
The complete obedience of any order
comes to the price of the obedient

The source that threatened my existence,
those that catalyzed the anger
Are the origin of the power that leads to their own
destruction
Dependence equal to air and water, my need for rage
becomes insatiable
How does it feel
To find yourself facing a foe that you created
You trained me with discipline and heart
You taught me to find something to fight for and I did
The right for all to live

No longer able to control the intensity, timing or
  direction
Becomes the cost for the advantages that once
  protected my soul
My temporary shield became the walls I carry today
Beware the utility of rage, although its uses may
  seem imperative
The cost may become too great when that debt is called
This is the cost of Army Strong
I would have rather been wrong
Hate filling the tanks that were made to burn off of
  love
the need to keep moving, carry on and move out
Never again will I need to get anywhere with such
  urgency
Looking for humanity in a place where there is none

I inhale the Love,
Pause to let the power seep and solidify into my heart
a sharp pain that is channeled and repurposed
I exhale a stronger, less brittle man

TO REMEMBER IS TO LIVE

Every step looking down I see the sand of a far away place
Giving me the feeling of traveling in time
Every loud sudden noise I turn to, as if my life is about
  to end
Giving me a new life every time
Every night I lock the doors, only to delay the danger
  that lives outside
Searching to erase my worry lines and find needed sleep
Every morning I look both ways crossing the threshold
  of my door
Giving me a view of the world in all its beauty
Every news report reminds me that the war is still alive
Giving me a mission to warn the youth
Every glass of wine is a celebration of survival
Opening to me a vision of those who didn't survive
Every bowl is a meditation in the explicit wonder of the
  Kingdom
Giving me a sense of home
Every friend reminds me I'm not alone
Giving me a Family that I can call my own
Every smile is a victory
Contagious to faces everywhere
Every kid I warn of war listens
Giving me purpose
Every piece of paper I recycle
Giving me the Jazz
Every step I walk is towards peace
Everyday I remember to live

ODYSSEUS

I have been home from combat for ten years
Ten years is how long it took Odysseus to return home
Ten years his beloved Penelope wove and unwove the
 same tapestry
Like a medic emptying and refilling aid station medical
 supply chests
Our survival requires us to think in a way that precludes
 learning
I got stuck in this way of thinking for years
Did you learn to learn again?
Did you study war?
Did you study other things?
What did you learn?
You learn what you live

GRIEVANCE LETTER TO DISCARDED UNIFORM

Dear Discarded Uniform,

I am sorry I have to destroy you. I am also sorry your owner gave you away but rest assured, you serve a higher purpose. While you will no longer clothe anyone, preventing sunburn from the desert sun or shielding from typhoon rains, you will be the object of creation.

You have seen many hard days; your owner might have worn you while he filled sandbags to repair a bunker. Your material absorbed his sweat, then held it to cool his skin. He wore you for days, maybe even weeks till he decided to wash you. Maybe he was the stinky soldier though, who thought your stench helped to ward off a pestering NCO.

Your pattern, your patches, your buttons, and limb holes all served a purpose and for this, I thank you. Stand detached now and destroy your form. Be reborn into a medium that will communicate the burden of your life and the one who wore you.

Respectfully I remain sincerely yours,

Alexander Fenno

HOW TO MAKE A COMBAT PAPER BOOK

Inspired by Chris Arendt's How To Make Combat Paper

1. Play Army in the woods
2. Put up F-14 Tomcat Jet-Fighter wallpaper above your bunk bed
3. Agree to a pizza date with the local Army recruiter
4. Graduate high school, watch the planes hit the towers, graduate basic training
5. mix 1 part nationalism with 1 part college money, stir in ½ baked optimism
6. Train, get desert gear, deploy to Iraq
7. Arrive in Kuwait, breath fumes from oil wells
8. Drive to Baghdad, load munitions onto truck, repeat for 3 months
9. Get flat tires, stares from Iraqis and meet friendly kids
10. Forget to strap down box of hand grenades, take a turn too fast, spill onto busy street, keep driving
11. Take pictures, don't change clothes, eat meals out of metal pouches
12. Watch traffic accidents, watch the truck in front of you burn, watch commanders get blown off burning truck by mortar rounds
13. Return home, get drunk, grind kitty litter into oil stains in motorpool, repeat for 3 months
14. Get out of the Army, enroll in college, get a job

15. Think about steps 1-13 often

16. Start writing, in groups, alone, in public, in the basement, repeat for 6 years

17. Read WWI poets, read Vietnam War poets, read Iraq War poets, become inspired by peers of the past and present

18. Collaborate, Collaborate, Collaborate

19. Start your first book with a poem about shitting in the sand

20. Send book to Harvard and your Grandparents

21. Ask for help, receive it, be grateful, live simple, speak your mind, plant seeds

22. Help others with steps 14-21, feel good again

DAVID KEEFE

My work explores and exposes the boundaries between reality and memory, between chronologically lived experiences and simultaneity. Fishing as a young boy and serving a tour of combat duty in Iraq converge inexplicably. The icy platform of Minnesota fuses with the ruin-dotted deserts of the Middle East. Fish become mortars and mortars become fish. The landscape is firm yet blurred and dream-like. These juxtapositions converge in my oil paintings and on handmade paper made from the uniforms I wore in Iraq, creating a visual conversation that can begin to inform the viewer of the continuity in life experiences, no matter how different.

REFLECTIONS AT BROWN'S POND

The morning sun shimmers across the pond,
a rippling glare that blinds us. We paddle,
our steady clap against the water's scum.
Bullfrogs dive away from the anchor's splash.
Witt and I prepare in private. We test drag,
the chink of hooks as we sort through old tackle.
We each have our own knots, our own way to deal
with tangles—he cuts his, I tease mine out.

We fish in silence, but some days, Witt begins
humming a cadence to which we both marched:
 a hymn
we share, whose verses tell of the killing we don't.
Reeling through the weeds, one of us will turn
to ask how the other is. But neither
ever says what he knows—just answers, Fine.

JOSHUA'S SONG

If the wind picks up out of the West
And shouts are heard on High.
My heart will have sunk within my chest
For dreams missed only in life.

The children will come to the noise
Of their Father's going.
Me, I cannot bear the cries: "There, look!"
Of silhouettes fast approaching.

Last times together are short.
Followed by distances vast and still.
Agape, ride fast into the Night!
Hope? If it is in your will!

A sky of purple takes you home!
An omen, is this true? I will do as can
Be done. Wearing your garland proud:
Goodbye. Until I see You.

TEN YEARS GONE

Ten years gone, 19 years old and naive
back flat against red Georgia clay-dirt
which wicks the sweat from my Army P.T. shirt
legs at roughly a 45 degree angle from the ground
hands to my sides, or tucked beneath my back
to protect a sore that had formed over my tailbone
teeth clenched and lightly caked with dust
eyes shut tight to avoid the relentless southern sun
and the frequent streams of sweat that run down my face
into a pool forming at the base of my neck
mind still full of the bullshit one believes
prior to knowing much of anything
unaware that by nine years gone
a different soil would claim my water

Nine years gone, 20 years old and a bigot
seated on the gunner's strap of a soft-skinned HMMWV
in Baghdad Iraq, the air apparent
radiating off buff colored buildings and yellowed Mercedes
fenders, quarter panels, and doors painted odd colors
or as often, simply removed
my whole world seemed caked with dust
fearful gaze fixed fast on anything and everything but
frequent streams of five-five-six could make me scream
fuck no, I'm not afraid!
mind so full of the hatred one acquires
following fear and death and failure
praying that by eight years gone
a different soil might claim my water

Eight years gone, 21 years old and tired
lead foot flat against that Chevrolet floorboard
attempting to feel something worthwhile
something but the numbness and the nothing,
 just this nothing
and the thought that all of life's pleasures and pains
may rightfully be retained from someone like me
ever recollecting how heinous I could be
knowing of a part of me I'd rather not have seen so
frequent dreams of who I'd been came crashing back at
 me
and into the hollow forming at the core of me
mind now full of the resentment one compounds
following loss without grief or healing
without concern for "life goes on"
that the earth continues to claim our water

My younger years gone, years older, still alive
feet feeling the sand, wading into warm water
fishing pole over my shoulder on a lake in fair Michigan
wind at my back, sweet scent of honeysuckle
my line flies fine, shot fired with great accuracy
the tug and the feel of the lure as I reel, heals me
eyes silently searching for the eddies and stills
 for my next opportunity
I frequent streams, the calm it brings makes me sing
fuck no, I'm not afraid
mind now equipped with the wisdom one acquires
having survived the times that try you
with the utmost concern that life should go on
that the earth may continue to cleanse with her water

THIS GROUND

Veterans Writing Group, Sebastopol, 2013

This ground alive with dead leaves underfoot,
    that holds the flower fast to the grave,

this ground of soil-flesh we are given to lose,
    that grieves me down the river and sings,

this ground I lay down on and listen to within,
    an earth-breath I think I hear,

this ground hollowed out and dried for a spark,
    would-be burning, green stalk dying,

    this is what we have to live for. . .

SUN SET

Thick pulses splattered vivid
red hues, or sullen orange gristle cast
on frothing clouds—like that foam
that gurgle-whistled out—a dying
wind: swayed through strands
of tall yellow grass bent, whispered
to broken damned things,
and hands that shake—that final glimpse
as the sun bled out, where shadows
stretched, and touched
and there was no end

Photo by Lovella Calica

THE MISSING LINK

"I practice being myself, and I have found
parts of myself never dreamed of by me."
−Jimmy Santiago Baca

A bold and deep quote that I have used as my internal compass since I heard it. Something I may not have ever read or been ready for at all, if not for the veterans' arts programs I've participated in, Warrior Writers especially.

I, like many combat veterans, came back from deployment "perfectly fine." That's what we say so the battalion will let us go see our families after being away for the most life-changing months of our lives. Lying to myself and everyone else that I was *fine* slowly ate away at me. As life piled more rocks in my pack it compressed the combat experiences down inside of me like old nitroglycerine, making me unstable and prone to eruptions that were changing me faster than I realized. Then it went off. It didn't cause a massive eruption that landed me on the 6 o'clock news. Instead, I imploded, like a sink hole of depression and guilt. I piled more responsibilities on myself to try to retain some self-worth.

Struggling as a single parent, I finally broke down and went for help. Weekly visits to Fort Belvoir's Behavioral Health did not go much deeper, at first, than the lie I had told everyone else about combat trauma. After all, I was *perfectly fine.* I had "bitten off too much," and I was "spread too thin." That was my excuse for why I was in the office.

As I whizzed through the psychologist's stress management tools and organizational life books, I started to realize how much I hurt inside. I asked for a referral to *Outpatient*. It only took one session of me being honest and explaining I wasn't *fine* to find myself as a patient of the *Co-occurring Program*. I still wasn't able to really grasp the concept of the problem though. At first I would go into a very controlled, clinical environment of people who were in varying degrees of *perfectly fine*. We would share experiences, thoughts and occasionally warn each other about which prescriptions to avoid. It would not be enough to drag me out of myself. I participated, and I talked, and gave advice, and it was spinning my wheels. I would dig up all this stuff and then go home where it would just sit on me. I had no outlet from 1630 to 0800 while outside of the clinic, but in my own head.

Then came the week when *Red Group* (combat trauma) went to writing group. That's when I got one of the most real pieces of advice I had ever heard: "Healing through writing is evidence-based when you dig in and you're honest with yourself. Write not just the *sitrep* version or the emotional version, but a combination of the two. It's gonna suck, you may feel like shit for a day or two, but once it's out, you will improve. That being said, if you need to talk to someone after this group, make sure you do, and be kind to yourself." That came from Seema Reza, and holy shit, it got my attention! So I listened to a poem by an artist I'd never heard of, an artist I previously would have written off and ignored. Then something happened. My pencil was moving as fast as my med-fueled brain and I related to the work. It opened up a little part of me that could be honest in a room full of strangers, that we were not *perfectly fine*.

That's when all the clinically-learned tools and lessons started to sink in. Not a lot, but a little bit. I went to my group sessions and my one-on-ones, and the internal gears were spinning enough to give me direction as to how I was going to whip this thing inside of me. Then came a workshop presented by Combat Paper NJ and Warrior Writers. That's the week when the steam really kicked in. The combination of Warrior Writers (using the same model of reading a poem, then writing in response to that poem), combined with the visual and physical process of hand-papermaking and printmaking, started to peel back layers of myself. The more it did, the better I felt, and the more creative I became.

I have been asked several times why I think these programs help so much. Here is my answer: It was a way to take my self-perception out of the picture and focus on something I felt people wanted to hear and see. It was different than giving a play-by-play of my *Alive Day* and telling a therapist I was on an emotional pain scale between *smiley face guy* and *grumpy face guy*. It gave *not being fine* a creative purpose and a way to be productive. Now it's not a pity party anymore. It's "Yes, I made this piece of art, yes this is my story, and yes I want to share it because I am a combat vet and

I was not fine...'til later." I still struggle and, for that reason, I make art.

Since that week in February 2013, I have been in four Combat Paper/Warrior Writers Workshops. In the last two, I participated as a facilitator. I still go to every writing group I can because they work. I am proof, and I've seen more first-hand from the community of veterans that has grown up out of these programs. From writing groups to words to song, Warrior Writers and Combat Paper offer the opportunities to perform and/or show your work. It builds that crucial link I didn't have before I got involved with art: the communication link.

If I can't communicate in a fire fight, I can't get support; if I can't get support, I can't move. If I can't move, eventually something heavy will drop with force and I will die. The same thing happens now if you take one of those links away: the communication link for trauma and emotions. The missing link.

Photo by Willie Young

AFTERWORD

During the past decade of warfare, approximately .5 percent of the American public has been active duty. Of those, more than 60 percent deployed at least once during that time, with many service members serving on multiple tours to combat zones. Veterans are a minority, and American attitudes concerning them tend toward passive gratitude, with occasional blooms of contempt and fear. To many, the armed forces represent uncomfortable truths that result from our decisions as a society. It is tempting and easy to turn away from their stories. Regardless of individual politics or personal connections to (or disconnection from) the military, it is all of our responsibility to bear witness with empathy to these narratives from the edge, to acknowledge the gray, and to offer acceptance to people whose service sometimes hinders their ability to love themselves. Veterans need access to communities of other veterans, certainly. But they also require blended communities that include civilians willing to extend the generosity of listening without judgment and the bravery of admitting ignorance and asking to understand.

My intention is not to induce shame over the decision of whether or not to serve in the military. I am asking that we think about what must be done to close the gap between civilians and service members because we need one another to establish a comprehensive account of our time. How can we create common ground and then move forward together to face the challenge of the broken world that was handed to us? I believe this collection of art and writing presents at least part of the solution.

Lovella Calica, founder and director of Warrior Writers, is not a veteran. Dedicated to this work for more than six years, she is a writer, a teacher, a bridge that sways but does not buckle. She is a force as real as gravity. Hers is the work of listening and guiding and inspiring the courage of looking inward. It is the work of a person compelled to understand what she has not had to experience, and driven to help those stories reach a broader audience. I have had the opportunity to accompany dozens of groups of Service Members to workshops facilitated by Lovella and her exceptional team of veteran writers. I have used the work from Warrior Writers anthologies in writing groups that I've facilitated myself, and every single time, I witness change and a greater sense of clarity—both in myself and in participants.

My first experience with Warrior Writers came in May of 2012. I accompanied a group of 12 Service Members receiving intensive mental health treatment at a military hospital where I work to an arts event at Corcoran University. We sat in a circle of desks in an empty basement classroom. Lovella directed the group to free write for a few minutes while she handed out copies of *After Action Review*, the organization's third anthology. There was a lot of tapping, a lot of scratching out, a lot of uncertain hovering of pen over paper. After a few minutes, we read "A Year of Secrets," a poem by Iris M. Feliciano in the anthology, aloud. Lovella then invited us to choose a personally significant date and write. This time the sound of the pens moving was steady. The resultant writing was stunning, clear, direct. Such is the power of art, and particularly the power of the anthology you hold in your hands. This work emerges from settings in which individuals have begun to make sense of the incomprehensible and begs to be used to create more such settings. Wherever you are, whoever you are— veteran or civilian—I hope you are moved to share it.

BIOGRAPHIES

Tom Aikens is a former 11B. He served as a team leader and a squad leader on deployments to Kosovo in 2003 and Iraq from 2004-2005. He is currently an assistant project manager with a construction company.

Michael Anthony served for six years in the Army Reserves and was deployed to Iraq. He is the author of the acclaimed war memoir, *Mass Casualties: A Young Medic's True Story of Death, Deception, and Dishonor in Iraq*, and is the War and Veterans editor for The Good Men Project blog.

Michael Applegate grew up in NJ. He served in the Navy from 1998-2006 as a cryptologist in Diego Garcia B.I.O.T. and Denver, CO, and as a submariner in Pearl Harbor, HI. He will begin graduate school in the fall at the School of the Art Institute in Chicago for Art Therapy.

Alan Asselin was raised in New Hampshire. He started writing poetry and designing houses when he was 13. He joined the Air Force in 1971 and almost immediately joined protests against the Vietnam War. He was discharged in 1972 and raised a family, mostly in Vermont. Alan spent several years incarcerated and has been in Boston since his release in 2006. He has been unemployed since 2009 and homeless for the last year.

Jan Barry is a Vietnam veteran, poet, and writer. He is the author of *Life After War & Other Poems* (Combat Paper Press) and co-editor of *Winning Hearts & Minds: War Poems by Vietnam Veterans*, among other works. He is active with Combat Paper and Warrior Writers.

Kevin Basl served two tours in Iraq with the Army as a mobile radar operator. He earned an MFA in fiction from Temple University, where he has taught in the first year writing program. He is currently a workshop facilitator with Warrior Writers and Combat Paper NJ.

Chantelle Bateman is a former Marine and longtime writer who has found Warrior Writers to be an important part of her healing process. She served in Al Asad, Iraq with MAG16 Forward from 2004-2005. She currently serves on the Board of Directors for Iraq Veterans Against the War (IVAW), working to help veterans unpack their military experiences and build healing communities working for change.

Greg Broseus is an OIF III combat veteran. He currently lives and works in Jackson, WY and is finishing his undergraduate thesis for his BFA at the School of the Art Institute of Chicago. He has exhibited at DiverseArts Culture Works, the National Veterans Art Museum, Mana Contemporary and Johalla Projects.

Michael Callahan served in the Marine Corps and the Army from 2003-2010. He deployed twice in support of Operation Iraqi Freedom (OIF). Originally from Peapack-Gladstone, NJ, Michael currently lives in Philadelphia. He is pursuing a graduate degree in social work and works for Combat Paper NJ as a Development Coordinator.

Drew Cameron is a second-generation hand papermaker, trained forester and former Army soldier. He co-founded the Combat Paper Project in 2007, where he facilitates workshops with veterans and the community transforming military uniforms into handmade paper, prints, and books. He is based in San Francisco at Shotwell Paper Mill and continues to practice papermaking, teaching and encouraging others to do the same.

Sean Casey served five years on active duty in the Army. He is currently an Army Reservist and graduate student studying public communication at Drexel University. He looks forward to a career in public relations.

Martin J. Cervantez is an Army artist still serving on active duty. He served as the artist in residence at the U.S. Army Center of Military History. A self-taught artist since childhood, Martin also does abstract works of art on his off time and employs his camera skills to capture everyday life wherever he travels.

Justin C. Cliburn served in Iraq with the Oklahoma Army National Guard in 2006. He was a Humvee gunner in a squad that traveled throughout Baghdad each day. In 2014, he graduated from the University of Oklahoma College of Law. He is married to the love of his life and lives and works in Oklahoma City.

Graham Clumpner served in the Army with the 2/75 Ranger Regiment from 2004-2007. He deployed for Operation Enduring Freedom (OEF) in Afghanistan and Pakistan. He believes that "People should not be afraid of their governments. Governments should be afraid of their people."

Jenn Cole served in the Air Force for four years. She deployed to Iraq in 2003 with the 728th Air Control Squadron. She currently lives in Cortland, NY with her wife. She utilizes writing, and arts and crafts to help cope with her war experiences.

David Connolly served with the 11th Armored Cavalry in Vietnam. He is proud to call himself a Vietnam Veteran Against the War and member of the Smedley D. Butler Brigade of Veterans For Peace (VFP). His collection of poetry and prose, *Lost in America,* was published by Viet Nam Generation & Burning Cities Press in 1994. David is a founder of the South Boston Arts Association and the South Boston Literary Gazette.

Tim Corrigan was a Medic with the Navy and Marines. Originally from the less-than-rugged lands of "Shaolin" (representing Wu-Tang), he is active with Warrior Writers

Boston, VFP, the Catholic Worker community, South End Tech Center, and Young Jewish Voice for Peace. He has spent over five years helping raise two young Iraqi refugees injured in the war, and is currently on leave from Stanford University.

Eric Daniels has been a Marine for nine years. He deployed twice with Lima Company, 3rd Battalion 6th Marines, 2nd Marine Division to Marjah, Afghanistan. He works at Officer Candidates School and studies at Northern Virginia Community College. His awards include the Purple Heart and Combat Action Ribbon.

Michael Day is a Marine combat veteran. He lives in Manhattan and is a cinematographer, producer, and writer for Cyprian Films NY. Last year he released his first book, *I am the Title*.

Maurice Emerson Decaul, a former Marine, is a poet, essayist, and playwright, whose work has been featured in *The New York Times*, *The Daily Beast* and others. His poems have been translated into French and Arabic and his plays have been performed in New York, Washington, DC and Paris. He is a graduate of Columbia University and is currently working towards his MFA at NYU.

Patrick Majid Doherty served in the Army and was discharged at the rank of Private after over three years of service. He was stationed in Manheim for 30 days and was court martialed. His discharge was later upgraded to honorable. Patrick studied writing at the William Joiner Center for the Study of War and Social Consequences at the University of Massachusetts Boston.

Christopher Duncan served as an Army infantryman. He deployed to Iraq for the first time as a shooter and again as the spotter of a sniper team. It was very hard for him to talk about the deployments and he was extremely self-reserved. He found relief from the inner demons through

Warrior Writers and Combat Paper NJ. He was medically retired as a Sergeant in April 2014.

W. D. Ehrhart is a Marine veteran of the Vietnam War. He is the author or editor of 21 books of poetry and prose and teaches English and history at the Haverford School in suburban Philadelphia.

Juston Eivers served as a nuclear machinist mate in the Navy on the USS Eisenhower from 1992-1996. He is extremely angered by the disservice done to combat veterans in the ongoing wars. Juston hopes Warrior Writers and other programs will help more veterans and involve more people in helping veterans.

Alexander Fenno retired from active duty with the Marines in 2014. He served in Iraq, Afghanistan, Cameroon, Nigeria, and Japan as a Bulk Fuel Specialist, Marine Security Guard, and Information Security Technician. He currently lives in Silver Spring, MD with his wife and kids. He enjoys taking long walks on the beach and reading conspiracy theories.

Stephen Funk joined the Marines in 2002 and declared himself a Conscientious Objector on April 1, 2003, becoming the first service-member to publicly refuse deployment to Iraq. He also came out publicly as a gay man at that time. He was court-martialed and served five months in the Brig. Stephen is an activist and the Founder and Artistic Director of Veteran Artists.

Shawn Ganther served in the Air Force from March 1998 through September 2002. He completed basic and technical training at Lackland AFB, TX. He was then assigned to the 375th Security Forces Squadron, Scott AFB, IL and deployed to Saudi Arabia, Kuwait, and Qatar.

Lindsay Gargotto is an Air Force veteran. She served on active duty from 2000-2004, where she was stationed at Elmendorf Air Force Base, AK and Lackland Air Force

Base, TX. She is a military sexual trauma survivor and warrior. She is now a cultural organizer and works for healing and justice for women veterans and active duty service members.

Adam M. Graaf served nine years in the Army Reserve and was deployed to Kuwait/Iraq in 2003. He is an MFA candidate at the University of Massachusetts Boston.

John Grote has served over 33 years in the military and is still serving on active duty at Fort Detrick, MD. He participated in a workshop with Combat Paper NJ and Warrior Writers at Fort Belvoir in 2013. John has had two sons deploy in support of OIF and OEF. He hopes to retire someday.

Jason Gunn spent 10 years with the Army and served in Iraq as a Tanker/Scout with the 1st Armored Division. He deployed to Iraq in May 2003. In November 2003, he was wounded after driving over a 155mm artillery shell with his HMMWV. After three months of recovery, he was redeployed back to Iraq where he took part in the siege of Karbala. He enjoys writing military science fiction and cartooning.

Toby Hartbarger joined the Army in high school. Within a year of graduating from high school, he was choking down sand in the desert in Iraq. He served one 15-month tour in Najaf, Sadr City, Baghdad, Fallujah, and Al Kut. He was trained as an Infantry Mortarman but served in Iraq as a scout. He is studying Biology as an undergraduate at the University of the Sciences Philadelphia.

Ryan Holleran enlisted as an infantryman in 2010. He was deployed for 11 months to Iraq in 2011. He spoke out against militarism while on active duty through Under the Hood and IVAW. Ryan serves on the Advisory Board for the Under the Hood Cafe and Outreach Center. His work can be found at ReactToContact.com.

Preston Hood served with SEAL TEAM 2 in Vietnam in 1970. His poetry has appeared in *The Café Review* and *Michigan Quarterly Review*. He recently published essays in *Prairie Schooner* and on The Stress to Success Factor blog. His book, *The Hallelujah of Listening,* published Červená Barva Press in 2011, won the 2012 Maine Literary Award for Poetry. He lives with his spouse, Barbara J. Noone, in Lyman, ME.

Aaron Hughes was called to active duty with the Illinois Army National Guard in 2003. He deployed to Kuwait and Iraq with the 1244th Transportation Company to haul supplies from bases in Kuwait to camps in Iraq. After three extensions, which resulted in a 15-month deployment, Aaron returned to the States with a vision to use art as a force to counter destruction and as a tool to confront militarism.

Andy Kaufmann was medically retired in 2009 after 23 years in the Army following multiple surgeries related to an incident in Iraq in 2004. He lives in the woods of Virginia with his family and dogs. He deals with PTSD and pain by painting and writing poetry.

David Keefe is an artist and the Director of Combat Paper NJ, a program of the Printmaking Center of NJ. Dave served as an Infantry Scout in the Marines and deployed to Iraq in 2006. His studio is in Montclair, NJ. Dave holds an MFA from Montclair State University, where he has taught part-time for four years.

Kevin Kilgore is a cartoonist, former Marine, and graduate of the Center for Cartoon Studies. He worked at Sangmyung University as a lecturer in the Department of Cartoon and Digital Contents and as an English teacher. His work has been published in *Leatherneck* magazine since 1997. He appeared in the *Chosun Ilbo*, *The Korea Times,* and *The Dallas Morning News.*

Ash Kyrie was raised in a blue-collar town in northern Wisconsin where he joined the Army National Guard at the age of 18. Ash was deployed to Iraq during OIF I in 2003. After returning from the war, he continued his studies and in 2007 he graduated with a BFA emphasizing sculpture and photography. He went on to acquire his MFA while minoring in Comparative Cultural Studies.

Zach LaPorte served in Iraq in 2005 and 2006 with the 2nd Ranger Battalion. After returning from war, he graduated from University of Wisconsin Madison with a bachelor's degree in Engineering Mechanics. He currently works as a mechanical engineer in the Chicago area and volunteers with veterans groups in the metro area.

Nathan Lewis joined the Army straight out of high school. September 11, 2001 was his second day of boot camp. He deployed to Iraq in 2003. He played a role of a U.S. soldier in *Green Zone,* a movie about the Iraq War. Nathan is the author of two poetry collections published by Combat Paper Press entitled *I Hacky Sacked in Iraq* (2009) and *Colors of Trees We Couldn't Name* (2012).

Elaine Little served as an Army interrogator in Afghanistan and as a broadcast journalist in Cuba and Bosnia. She is currently working on a novel about the mobilization of a group of Army Reservists and feels the wartime experience, particularly from the female soldier's perspective, has yet to be satisfactorily explored. Her work was also published in the acclaimed anthology, *Powder: Writing by Women in the Ranks from Vietnam to Iraq.*

Iris Madelyn is a writer and artist from Chicago. She served in the Marines from 1996-2010 and deployed in support of OEF in 2002. Iris advocates for the therapeutic effects of the arts and focuses her work on creative expression as a means for social consciousness and healing.

Fred Marchant was honorably discharged from the Marine Corps as a conscientious objector in 1970. His book *Tipping Point* won the 1993 Washington Prize and was re-issued recently in a 20th Anniversary Second Edition. He is also the author of *Full Moon Boat* and *The Looking House*, both from Graywolf Press.

Hugh Martin is a veteran of the Iraq War and the author of *The Stick Soldiers* (BOA Editions, Ltd., 2013) and *So, How Was the War?* (Kent State UP, 2010). His work has recently appeared in *The Kenyon Review*, *The New Republic*, and *The New York Times'* At War blog. Martin has an MFA from Arizona State and he is currently a Stegner Fellow at Stanford University.

Maggie Martin served in the Army from 2001-2006. She deployed three times to Kuwait and Iraq. She has been with Warrior Writers since 2008 and works as an organizer with IVAW.

Rachel McNeill joined the Army Reserves as a heavy construction equipment operator in 2002 at age 17. She deployed to Ramadi, Iraq with the 983rd Engineer Battalion from 2004-2005 where she served on a convoy security team. She was medically retired in 2010 and graduated *cum laude* with an ALB in Extension Studies from Harvard University in 2014.

Joe Merritt enlisted in the Marines in 2006. He spent four years with 3rd Battalion 8th Marines as a machine gunner in a Combined Anti Armor Team Platoon. He was stationed at MCB Quantico in 2010, where he taught Machineguns at T.B.S. Joe was introduced to Combat Paper NJ and Warrior Writers in February 2013 and has attended several workshops and conferences since. He is a facilitator with CPNJ.

Sarah N. Mess served on active duty in the Army from 1992-2000. In 1993 she deployed to Mogadishu, Somalia with the 42nd Field Hospital as a 91D Operating Room Specialist in support of Operation Restore Hope and later Operation Continue Hope. With very few boots on the ground her role was expanded to include combat roles. She currently resides in New Jersey with her husband and their two children.

Jason Mizula served in the Coast Guard and Army National Guard from 2002-2008. He was deployed to New Orleans after Katrina, in the Caribbean and East Pacific on counter-narcotics and immigration patrols, and to Iraq during the surge. He attended college in Boston, where he was arrested in 2011 for civil disobedience alongside fellow members of VFP. Jason currently lives off the grid in Hawai'i with a few dozen chickens.

Brandon Moore-McNew served as a military policeman in the Missouri Army National Guard. He was deployed twice to Afghanistan in addition to participating in stateside service. He currently lives in Virginia, where he studies poetry.

Nick Morgan served as an engineer in the Army Reserves from 2002-2006. He served in Baghdad, Iraq during OIF II in A Company, 458th Engineer Battalion under the 1st Cavalry Division. He lives in Colorado and is finishing his bachelor's degree in ecology and evolutionary biology.

Robynn Murray is a very low-key writer and parent spending much time with close friends and family. She served in the Army during wartime, which is often referenced in her writing, art, and activism. She currently speaks all over the country and world about PTSD and veterans issues. She was the subject of the 2011 Oscar-nominated documentary *Poster Girl*. She likes punk rock, civil disobedience and long walks on the beach.

Walt Nygard is a Marine Corps veteran of the Vietnam War. He was born in Portland, OR and raised on Army posts of the American West and Germany. Walt is a blue-collar writer, the author of *The Summer Joe Joined the Army* (Post Traumatic Press, 2010) and a paper/print-maker working with Combat Paper NJ. His oldest son, Joe, served in Afghanistan and Iraq with the 10th Mountain Division.

Jennifer Pacanowski served in Iraq from 2004-2005 as a combat medic on convoys with the Army and Marines. She was diagnosed with PTSD in 2007. She writes to sort through issues she dealt with after war and finds solace fostering bullmastiffs. She found peace in the company of veterans and a venue for her art through Warrior Writers. She is a public speaker, poet, writing facilitator, and volunteers for Veterans' Sanctuary in Ithaca, NY.

Giuseppe Pellicano served as a Medic in the Army from 2000-2004 in Germany and California with a deployment to Kosovo. He received his BA in Studio Arts from North Central College in Naperville, IL in 2012. Giuseppe aims to raise awareness about humanitarian issues as well as the military experience. He works in various mediums, including photography, ceramics, and metals.

Garett Reppenhagen served as a cavalry scout sniper with the Army's 1st Infantry Division in Kosovo and Iraq. Garett co-authored the anti-war blog Fight To Survive while still serving on active duty in Iraq. Garett currently works as a social justice organizer and a veteran advocate in Denver, CO.

Benjamin Schrader served as a Cavalry Scout and deployed to Iraq from 2004-2005. Since exiting the military, he has been focused on school and is currently working on his PhD in Political Science at the University of Hawai'i. His focus is on war, veterans, social justice, and activism.

Rina Shah is a medically retired Army Captain. She served in combat in Iraq as a JAG Officer. She attended workshops with Combat Paper NJ and Warrior Writers in Washington, DC in 2013 and 2014. She is the Founder and Executive Director of International Spirit of Healing, a non-profit organization supporting military and veteran families.

Krista Shultz served as an interrogator and Arabic linguist in the Army from 1988-1991. She was in Saudi Arabia and Iraq for Operation Desert Shield/Operation Desert Storm. When she left the military, she had PTSD without realizing it. Several years ago she met a psychologist who worked with veterans with PTSD and began her journey towards healing. Writing poems about her experiences is one way she finds healing.

Erica Slone was born in Cambridge, OH. She served in the Air Force and deployed multiple times in support of the Global War on Terror. She received her Bachelor of Fine Arts degree from Ohio State University and is currently an artist-in-residence at the National Veterans Art Museum in Chicago, IL.

Kimberly Smith was deployed from May 2009 to January 2010 with 1st Battalion 5th Marines to Helmand Province, Afghanistan. She was the first female to deploy with an infantry battalion. Captain Smith is currently with NICoE at Ft. Belvoir and resides at the Fisher House. She is receiving treatment for severe complex PTSD and moderate TBI. A stroke and seizures caused loss of mobility and strength due to injuries sustained in combat.

Michael Spinnato served in the Marine Corps from 2003-2007 because of 9/11, believing this was the best way to respond to a humanitarian crisis. He is now going to medical school because he does not want to carry aid in one hand and a rifle in the other. As a physician, he will work for global health equity and social justice. Michael helped found Warrior Writers in Boston in 2013.

Jeremy Stainthorp Berggren is is a writer, poet, and visual artist. He served in the Marine Corps from 1998-2004. Jeremy has been involved in Warrior Writers as a participant, performer, organizer, and facilitator. He is involved with IVAW and various arts and cultural activities. He is included in *Truth Underground*, an upcoming documentary about spoken word artists in the triangle area of North Carolina.

Lamont B. Steptoe is a Vietnam veteran who served with the 38th Scout Dog Platoon of the 25th Infantry Division in Cu Chi, Vietnam. He is from Pittsburgh, PA, and is the author of three collections of poems about his time in Vietnan: *Mad Minute*, *Dusty Road*, and *Uncle's South China Sea Blue Nightmare*. He completed 11 other collections and has received an American Book Award and a Pew Fellowship in the Arts.

Peter Sullivan served in the Army National Guard for 12 years as an infantry soldier. He has been drawing and making art since he was about five years old.

Brian Turner earned an MFA before serving for seven years in the Army. He was an infantry team leader for a year in Iraq beginning November 2003 with the 3rd Stryker Brigade Combat Team, 2nd Infantry Division. His poetry has been published in many journals and in the *Voices in Wartime Anthology* in conjunction with a feature-length documentary of the same name. He received a 2007 NEA Literature Fellowship in poetry.

Jon Turner served two combat tours in Iraq and a humanitarian mission to Haiti as an infantryman with the Marines. He returned to utilize creativity to understand his time in conflict and as a method for re-integration. He recently opened the Center for Healthy Change and received certificates in Sustainable Building and Design, Permaculture Design, and Natural Building. He currently lives in Vermont with his family and animals.

Hart Viges joined the Army in reaction to 9/11. Hart served with the 82nd Airborne Division 1/325 as a mortar man in Iraq from 2003-2004. After returning, he filed for CO status and received an honorable discharge. He continues his objection with IVAW and VFP. Based in Austin, TX, he's focused on Operation SOY (Sustainable Options for Youth) where popular education meets the militarized youth.

Eric Wasileski (M. Div.) is a Persian Gulf veteran of Operation Desert Fox and of former Yugoslavia. He is writing a meditation book for veterans, *Handbook for Returning Veterans and Those who Love Us*, expected to be out in the spring. He is a Quaker and active with VFP and Warrior Writers Boston. He recently founded Veteran-Friends, a Quaker veterans group.

Paul Wasserman served in Iraq as an Army sergeant in a reconnaissance aircrew. He published a volume of poems, *Say Again All*, as well as a volume in the Veterans Book Project series. He lives in Brooklyn and works as a teacher.

Bruce Weigl enlisted in the Army and served in Vietnam for one year. Bruce is the author of more than a dozen books of poetry and has won numerous awards for his work, including the Robert Creeley Award, the Lannan Literary Award for Poetry, the Poet's Prize from the Academy of American Poets, and two Pushcart Prizes. He has received fellowships from the National Endowment for the Arts and the Yaddo Foundation.

Eli Wright served as a combat medic in the Army from 2002-2008. He was deployed to Ramadi, Iraq with the 1st Infantry Division from 2003-2004. After returning home and struggling to make sense of the experience, he began participating in Warrior Writers and the Combat Paper Project in 2007. He has been teaching papermaking and printmaking to veterans since leaving the military and is an instructor for Combat Paper NJ.

Emily Yates loves to write and always has. For six years, she was an Army "journalist" – notated as such because military journalism is called "public affairs" – and wrote many articles and editorials on topics revolving around the Army. She writes and performs smart-ass folk songs. She currently writes an advice column called "Ask Someone Who Knows Stuff About Things," which you can read at emilyyatesdoeseverything.com.

James Yee is a former Army Chaplain for the military prison at Guantanamo Bay. After objecting to the abuse of prisoners, he was arrested and sent to solitary confinement for 76 days. Charges were dropped soon after. He authored *For God and Country: Faith and Patriotism Under Fire*, which was featured in TIME Magazine's 9/11 tenth anniversary issue, and participated in the CNN/HBO documentary "Beyond 9/11: Portraits of Resilience."

Willie Young has served on active duty in U.S. Army Military Police Corps since May 1995. His service includes deployments in support of peacekeeping operations in Kosovo in 2000 and a combat tour in Afghanistan in support of OEF in 2004. Willie became a part of the Warrior Writers/Combat Paper NJ team following hospitalization for symptoms of PTSD. He is also a photographer.

Ty Zabel served with the Army National Guard for four years and did not deploy. His unit was 11 Bravo/Infantry. Tyler was a Specialist, but was bumped down to PV2 after going AWOL. He was released as a Conscientious Objector.

ACKNOWLEDGMENTS

We are grateful to the many people who helped make this book a success. Thank you to each and every veteran who submitted work and to their family and friends who continue to support them. We have loads of gratitude to our 112 Indiegogo campaign supporters, we couldn't have done it without you. Special thanks to: Jan Barry, Kevin Basl, Chantelle Bateman, Julie Batten, Roger Bonair-Agard, Kimberly Bonner, Lovella Calica, Combat Paper Project, Combat Paper NJ, Printmaking Center of New Jersey, Isabel Cook, Brandon D'Augustine, Michael Day, Caroline DeLuca, Lynn Estomin, Adam M. Graaf, Toby Hartbarger, Tony Heriza, Tyler Horst, Matt W. Howard, Aaron Hughes, John Kaeser, Yusef Komunyakaa, L. Brown and Sons, Iris Madelyn, Fred Marchant, Rachel McNeill, Ladan Osman, Seema Reza, Jimmy Santiago Baca, Carlos Sirah, Jeremy Stainthorp Berggren, Jason Tauches, Jon Turner, Michael Spinnato, Eli Wright, and Willie Young.

INDEX